INTERNAL CONFLICTS IN SOUTH ASIA

ᴹPRIO

International Peace Research Institute, Oslo
Fuglehauggata 11, N-0260 Oslo, Norway
Telephone: (47) 22 55 71 50
Telefax: (47) 22 55 84 22
Cable address: PEACERESEARCH OSLO
E-mail: info@prio.no

The International Peace Research Institute, Oslo (PRIO) is an independent international institute of peace and conflict research, founded in 1959. It is governed by an international Governing Board of seven individuals, and is financed mainly by the Norwegian Ministry for Education, Research, and Church Affairs. The results of all PRIO research are available to the public.

PRIO's publications include the quarterlies *Journal of Peace Research* (1964–) and *Security Dialogue* (formerly *Bulletin of Peace Proposal*, 1969–   ) and a series of books. Recent titles include:

Robert Bathurst: *Intelligence and the Mirror: On Creating an Enemy* (1993)

Nils Petter Gleditsch et al.: *The Wages of Peace: Disarmament in a Small Industrialized Economy* (1994)

Jørn Gjelstad & Olav Njølstad, eds: *Nuclear Rivalry and International Order* (1996)

INTERNAL CONFLICTS IN SOUTH ASIA

Edited by

KUMAR RUPESINGHE
and KHAWAR MUMTAZ

⚘PRIO

International Peace Research Institute, Oslo

SAGE Publications
London • Thousand Oaks • New Delhi

SAGE Publications Ltd
6 Bonhill Street
London EC2A 4PU

SAGE Publications Inc
2455 Teller Road
Thousand Oaks, CA 91320

SAGE Publications India Pvt Ltd
32, M-Block Market
Greater Kailash – I
New Delhi 110 048

British Library Cataloguing in Publication data
A catalogue record for this book is available from the British Library.

ISBN 0 8039 7752–2

Library of Congress catalog record available

Typeset by M Rules
Printed in Great Britain by Redwood Books, Trowbridge, Wiltshire

Contents

Foreword

The half-century-long enterprise of the governments of South Asia to form nations out of diverse societies and cultures has produced functioning parliamentary democracies, but the region is marked by war and violence. At the moment of Independence from British rule in 1947, the partition of India and Pakistan was accomplished amid terrible bloodshed and the forced mass migration of people. Since then there has been a succession of civil wars and three international wars. Recent years have seen new surges in communal tensions and violence. The total of those who have died in the region's wars or in war-related famines or who have been maimed or who have been displaced is more than ten million. Today, refugees and displaced people number well in excess of two million.

Regional conflict has both national and international dimensions. In India, the common thread of the civil wars in Jammu and Kashmir, Punjab, Assam and Uttar Pradesh is the rejection of the New Delhi government's authority. Pakistan has interspersed parliamentary government with longer periods of military rule and, though Pakistan is now under civilian government, the endemic violence of its politics means that military takeover remains a possibility. Bangladesh broke out of Pakistan in a war of extreme ferocity in 1972 and has endured lower levels of civil war almost constantly since then. Sri Lanka exploded into war in 1983 on two fronts: the government was challenged by a general guerrilla war launched by the revolutionary JVP, and by the secessionist war of the Liberation Tigers of Tamil Eelam in the north-eastern part of the country. The latter war was still going on at the time of writing.

These civil wars have been fought out against an international backdrop shaped by intense rivalry and confrontation between India and Pakistan. Both states are well armed: in the second half of the 1980s, India became and remains the largest Third World arms importer while Pakistan is the fifth largest, though with a volume of imports that is only about one third of India's. Both also have significant arms-manufacturing capacities and both are believed able to make nuclear weapons (India detonated a nuclear explosive in 1974), though neither has openly announced possession of them. The two states have gone to war three times since 1947, are formally still in a state of war with each other, and became involved in a display of confrontational brinkmanship in 1990 in which, on both sides, their nuclear weapons capability became an issue; though events are still shrouded in secrecy, it seems that there was a nuclear alert on both sides.

Under these wars and conflicts lie more levels of violence. Both random and organized crime are rising in several parts of the region, especially drug-trafficking and crimes related to it. One side-effect of the war in Afghanistan was to introduce extraordinary numbers of small arms into Pakistan, arming both its crime and its politics. In India, religious sentiment often leads to communal division, conflict and violence. There is a strong social basis for violence and conflict, despite some very important signs of economic development since Independence. The population of the region is growing faster than economic output, and there is large-scale migration from rural to urban areas, much of it as a result of severe depletion of resources and degradation of the natural environment. Though the survival of parliamentary democracy remains a cause of hope for peaceful resolution of some of the conflicts, democracy has suffered from systematic abuse by the power of both money and violence.

This volume has grown out of a long period of work at the International Peace Research Institute, Oslo, on the causes and dynamics of conflicts in Third World countries. This research is fundamentally shaped by a desire to identify possibilities for transforming and resolving conflicts. A main figure in the Institute's research on this theme has been Kumar Rupesinghe, one of the editors of this volume. His academic background is in sociology. His co-editor, Khawar Mumtaz, is an author and activist in Pakistan, perhaps best known for her writing on women's issues, with an academic background in international relations. The authors they have brought together for this anthology explain and illuminate the problems of the region, collectively providing a much-needed interdisciplinary overview of the nature and dynamics of the conflicts. The chapters offer an exploration of dimensions of regional politics and conflict that even the relatively well informed normally ignore, notably including the basis of the mobilization of women by the BJP in India and in the Sindh conflict in Pakistan. There is a wealth of empirical and historical information here that can form the necessary starting-point for thinking positively and creatively about the sources of solutions.

Dan Smith
Director, PRIO
Oslo
November 1995

Acknowledgements

The production of this volume has been a complex process coordinated by people on two continents and involving contributors scattered across South Asia, Europe and beyond. Special acknowledgement is due to the International Peace Research Institute, Oslo (PRIO), for supporting the initiative to get the dialogue started between scholars of the South Asia region, and for subsequently sustaining it, as well as to Halle Jørn Hanssen and Ann Bauer of the Norwegian Ministry of Foreign Affairs; to Smitu Kothari as one of the original conceptualizers of the project; to all those who acted as referees; and to the Marga Institute in Colombo. Those connected with PRIO in Oslo and Shirkat Gah in Lahore, particularly Jan Helge Hordnes, Shahid Iqbal, Nina Gera and Ambar Naveed, merit special mention for their skills, diligence and advice. We would also like to thank Dan Smith, the Director of PRIO, for his support; David Lord and David Reed at International Alert for coordinating the editing of the volume; and David Hill at our publishers, Sage, UK, for encouragement and sound advice during the editorial process. Our families, as usual, provided support on the home fronts, making possible what we have managed to achieve.

Kumar Rupesinghe
Khawar Mumtaz

Introduction

KUMAR RUPESINGHE AND KHAWAR MUMTAZ

The scale and intensity of violent internal conflicts in South Asia have esca-lated dramatically in recent decades. Long-running conflicts have become more deeply entrenched and newer ones appear to be taking on the intractability of their precursors.

Manifestations of that violent trend include the razing by militant Indian Hindus of the Ayodhya Mosque in December 1992, the consequent commu-nal violence in which an estimated 2,000 people died and the devastating terror bombings in Bombay and Calcutta in March 1993; violent clashes in Kashmir between Muslim militants and Indian security forces, in Punjab between Sikh militants and Indian forces; and the growth of militant auton-omist movements in Sindh and increasing lawlessness and religious sectarianism throughout Pakistan. Elsewhere in the region, repression of the Chittagong Hill Tribes in Bangladesh continues, as well as the brutal civil war and associated political violence in Sri Lanka.

The rising degree, multiplying varieties and the brutality of expressions of hatred have hardened positions and hearts throughout the region. And while there have been peaks and lulls in the violence, the underlying causes stubbornly persist.

Identity and the meaningful recognition of identity have been central to violent internal conflicts. Politically, these assertions have surfaced in demands ranging through degrees of greater autonomy, to federalism, to independence. Claims are being articulated as group rights, rather than in terms of individual rights. The final outcome of the collapse of the Soviet Union and the bloody disintegration of federal Yugoslavia has yet to be played out, but the implications of the shattering of the Soviet empire will reverberate throughout the world for decades to come, not least in the colonially constituted, ethnically diverse states of South Asia.

Many of the current South Asian conflicts share common characteristics, including fundamental questioning of the impact of creating or sustaining national states as the most effective framework for development when they include plural and socially differentiated societies. Rapid economic develop-ment and the incorporation of more and more people into the market economy, once seen as a process which would weaken more rigid traditional societies and dissipate the friction generated by the clash of the traditional with the modern, has also been brought into question. At the same time, the assumption that security and equality for minorities would be a natural outcome of integration into national societies has been shown to be seriously flawed.

Throughout the region, the processes of popular participation, particularly in electoral politics, have been abused and misused, widening existing fissures and creating new ones. While there is no doubt that electoral politics are an indispensable instrument for effective and equitable social organization, and in the short term can defuse potentially volatile social and political situations, in the South Asian context, where genuine democratization and decentralization of power have not taken root, elections have been consistently manipulated to consolidate the hegemony of dominant sectors within each country. Recent electoral setbacks for extremist opposition political groups in Pakistan and India at least afford some breathing space for possible reform, although there is little indication that leaders have the political will to attack the underlying causes of violence. In Pakistan the loss of several seats by the religious parties and the defeat of the pro-Hindu Bharatiya Janata Party in elections in northern India in late 1993 are likely to be only temporary setbacks for proponents of exclusion and intolerance, largely because South Asia's aspiring new elites continue to gravitate toward religious and nationalist parties which preach exclusionist policies and are in collision with the postcolonial succession elites – the secularized, Western-educated establishment.

Other potential defenders of civil society – the military and police – have been prime actors in the partisan and brutal repression of internal conflicts, which has been 'legitimized' by the promulgation of numerous laws that restrict individual and group rights and exacerbate tensions. State-controlled media have also played a role in the legitimization of repression. It is a sad irony that the growth of military and police establishments supported by high proportions of national budgets has not been able to check the erosion of the monopoly on violence of national governments. In fact, lawlessness, particularly in Pakistan and India, has combined with state-sanctioned violence and criminality to create cultures of violence and insecurity, with an upsurge in para-military groups, goondas, thugs, death squads and private armies.

The behaviour of governments, which were conceived as essentially secular and non-partisan mediators between various sectors of society, has degenerated to the point where state actions have increasingly contributed to social strife by bolstering dominant sectors of society and directly and indirectly repressing minority ethnic, religious or political groups. Adding to the already enormous tensions in the region has been the rapid growth of various forms of religious fundamentalism, in some cases state-sponsored as a means of co-opting segments of the population or nurturing a counterfoil to opponents of the state. In India, Pakistan and Sri Lanka, religion is at the centre of some of those societies' most severe conflicts – holding the promise of ongoing widespread sectarian violence, but also the potential for mobilizing the faithful along avenues of peace, reconciliation and healing.

The dominant models of development in South Asia have given rise to social tensions and conflict by generating new insecurities for the most vulnerable sectors of society. Examples of this process include the displacement of thousands of people, who are rarely adequately compensated, for large-scale industrial or agricultural projects such as the Mahaweli project in Sri Lanka, the Tarbela Dam in Pakistan and the dams on the Narmada River in India; the depletion of natural resources to fuel industrialization or export-led growth; import liberalization with negative impacts on the livelihoods and lifestyles of millions of people.

While South Asia was not free of conflict prior to the colonial period and the advent of industrialization, most historians agree that there was relatively harmonious coexistence between religious communities. The deep and pervasive hostility between Muslim and Hindu in India is a modern phenomenon; Sri Lanka (then Ceylon) was not the scene of the debilitating conflict in which Sinhalese and Tamils have been locked; the marginalization and exploitation of Sindhis in Pakistan by Punjabis is also a post-Independence (1947) development.

While this volume explores internal conflicts in the region, many of them under discussion have transnational or regional aspects – Sri Lankan Tamil links with Tamil Nadu, Indian Muslim links with Pakistan, Indian and Pakistani contention over Kashmir, which escalated dramatically in 1993, support by Sikhs in India for nationalist movements within the Punjab – the list is lengthy. Furthermore, international spillovers from internal conflicts could have disastrous effects in a region where the two main military powers are believed to possess nuclear weapons and sophisticated delivery systems, as well as large and well-equipped conventional forces.

The rich and varied collection of chapters contained in this volume illustrates these general trends and others. The chapters also provide numerous insights into the particularities of internal conflicts in India, Pakistan, Sri Lanka and Bangladesh.

Kumar David, in his chapter on *Ethnic Conflict: Rethinking the Fundamentals*, points out that ethnicity is at the core of many of the internal conflicts present in India and elsewhere, but rejects the notion that 'the world should return to the patchwork mosaic of . . . myriad kingdoms, tribes, principalities, states and fiefdoms, each in the modern sense a separate or semi-state'.

David argues that larger national units are better able to marshall and organize their resources and economic production than smaller entities; a mosaic of small nations with their barriers and exclusive symbols would be a hindrance to the full development of human potential; and the integration of smaller nations into larger national states reduces the scope for exploitation by more powerful imperial or neighbouring states. David contends that 'the ideology of ethnicity must be rejected as false consciousness' and that the 'economic unification of the world is irreversible'. He suggests a two-level approach to the negative aspects of ethno-politics: a commitment to oppose racial and religious oppression, as well as oppressive regimes and social orders; and a commitment to a new world order and a new universal human consciousness.

In contrast to David's world integrist scenario, **Dev Nathan**, in *India: From Civilization to Nations*, sets out a Marxian argument for ensuring the right of separation for minorities within India. Nathan contends that 'the attempt to fashion a homogeneous national culture out of a civilization has led to a clash of cultures and of languages' and the compulsions of development, progress and security in India are those of the 'pan-Indian big bourgeoisie'.

> They are not the compulsions of those who want to build a more democratic order. An essential component of such a democratic order is that there should be an end to national oppression, that there should be national equality.

However, Nathan recognizes that some of those seeking to wrest more economic space from the current Indian elite are motivated by the desire to 'create better conditions for their own accumulation'. Ultimately, he contends, the struggle for a more democratic society must be linked to ensuring the right of separation, as well as the economic and political rights of minorities.

Akmal Hussain traces the evolution of the relative powers of the military, the bureaucracy, political actors and the Pakistani population in *The Dynamics of Power: Military, Bureaucracy and the People*. Hussain describes the distorted economic growth of the country, its foreign indebtedness and dependence on aid, as well as the fragmentation of civil society caused by the widespread availability of surplus arms from the Afghan war, the growth of the heroin trade and the erosion of the state's monopoly on violence. In contrast to the decaying bureaucracy, the Pakistani military has emerged as a highly educated and professional institution which dominates the country's political life. However, the author notes that, on several occasions since Independence, the population has been able to retake the democratic terrain and press 'demands for social equality, justice and political representation of the dispossessed', thus remaining 'a factor to be reckoned with by those who pull the levers of power within the state structure'.

While the Islamic 'fundamentalist' movement contained the potential for rethinking the meaning of the Koran by Islamic scholars and contributing to a renaissance of Islam, **Abbas Rashid** notes in *Pakistan: the Politics of 'Fundamentalism'* that the revivalist effort was constrained by a 'traditionalist intellectual framework' and did not attempt to break those bonds by promoting challenge and inquiry.

The founders of Pakistan may have thought about the application of Islamic principles in the new state, Rashid notes, but they were essentially Westernized secularists, who treated the Jamaat-e-Islami and other Islamist parties as nuisances, to be placated with occasional concessions. That approach was dropped after March 1953, when the 'fundamentalists' mounted a strong challenge to government authority over the presence of Ahmadis (members of a sub-sect of Islam declared non-Muslim during General Zia-ul-Haq's rule), in key posts. The ensuing riots and the imposition of Martial Law set the pattern for succeeding governments to adopt 'an Islamic formalism to assuage what they saw as potentially disruptive forces in society'. The July 1977 coup that brought General Zia to power ushered in a new era of state-sponsored 'fundamentalism' with the inclusion of Jamaat-e-Islami ministers in the cabinet and the promulgation of constitutional changes and new criminal statutes which provided for punishments such as stoning for adultery, the amputation of a hand or foot for theft and flogging for false accusations of adultery or the consumption of alcohol.

Rashid asserts that while Islamic 'fundamentalists' do not command enough support within Pakistan or show enough internal cohesion to form a government, they will continue to have a strong influence over governments. It is also apparent that the tide is running against the evolution of an indigenous and authentic liberalism that could address Pakistan's democratic and developmental needs. The author suggests that reversing that tide will require the reinsertion of rationalists and humanists into the mainstream of Islamic thought. He also suggests that that can be accomplished only by decoupling the notions of Westernization and modernization.

In *The Politics of Violence in the Indian State and Society*, **Sumanta Banerjee** contends that India's internal sources of violence – secessionist terrorism and insurgencies, electoral violence, and violence plaguing Hindu–Muslim relations – are rooted in inequitable distribution of wealth and imbalances in regional growth. 'The perceived sense of injustice among the people is becoming fissured along the lines of caste, ethnic loyalties, language, religious beliefs – differences and divisions which the Indian polity has failed to overcome, despite its commitment to a "socialist secular democratic republic".' Governments have allowed issues to fester, acting – often with high levels of violence and brutality against innocent civilians – only when militant agitation grows. In this climate, anti-state mass movements, guerrilla movements and terrorism have emerged with a vengeance. In Punjab and Kashmir, nationalist terrorist groups are seeking to establish autocratic theocracies and have been targeting state authorities, leftist groups and leaders, and religious moderates. Violent state countermeasures, insomuch as they single out visible ordinary citizens rather than invisible terrorists, have alienated the population and filled the ranks of terrorist groups.

As to Kashmir, the author attributes the strength of the nationalist movement to the 'throttling of the normal democratic process' through the rigging of elections and the appointment of unpopular Chief Ministers by New Delhi. Banerjee also notes that the opponents of the state have graduated from vintage arms to state-of-the-art weapons. In general: 'The language of the gun is thus becoming a decisive force in political discourse in India.' Violence has also invaded the party political sphere since the assassination of Indira Gandhi and the subsequent massacre of an estimated 3,000 Sikhs in Delhi, allegedly orchestrated by Congress (I) party officials who were never prosecuted for their roles in the pogrom. The subsequent assassination of Rajiv Gandhi in 1991 highlighted 'the general culture of violence that marks Indian society today'. The 'criminalization of politics' – the acquisition of legitimacy by a 'vast underworld of smugglers, hired killers, gun runners and gangsters' used by politicians to influence elections – is another factor fanning the flames of communal violence.

Banerjee states that Indian society is faced with two options: disintegration, or dialogue, accommodation and the evolution of a new model for society. In the latter case, what is proposed is the evolution of a 'framework to harmonize these different identities that divide society today, at the same time allowing the peaceful development of each group'. Banerjee recommends revisiting the controversial proposal made in 1973 by the Sikh Shiromani Akali Dal to restructure the Indian state, confining the central government to responsibility for defence, currency, foreign relations and communications. A national dialogue based on such a proposal should include all the major actors for all sections of Indian society.

The growing resort to violence in Pakistan in conflicts involving political rivalries, disputes between student organizations or between tenants and landlords is the central theme of **Shireen M. Mazari**'s contribution: *Militarism and the Militarization of Pakistan's Civil Society: 1977–1990*. The imposition of martial law instituted after General Zia's power grab in July 1977 marked a watershed in the militarization of Pakistani society and the resurgence of the Pakistani military after its humiliation in the loss of East Pakistan and defeat by India in 1971. The re-emergence of the military, the use of terror to rule, the development of the arms and drug culture overflowing from the Afghan war, the stifling of political parties and dissent – all served to compound the pressures created by ethnic, sectarian and kinship cleavages in society.

In Sindh in the mid-1980s, private armies of tribal leaders were revived, political opponents of Zia joined with dacoits in armed gangs, and the general level of violence led to ethnic polarization. The establishment of the Movement for the Restoration of Democracy, which brought together the main national political parties, was soon followed by the setting up of the militant Mohajir Qaumi Mahaz or Mohajir Nationalist Movement, with its fascist structure and armed cadres. After more than a decade of military rule, the author notes, the militarization of society and the ongoing violence had acquired their own momentum – with peaks and lulls, to be sure. The death of Zia and the election of civilian governments have yet to reverse that momentum.

Meghna Guhathakurta examines the dynamics of democratization in Bangladesh that led to the ousting of General Ershad and the parliamentary elections of 1991 in *Democratization in Bangladesh: the Mass Uprising of 1990 and Its Aftermath*. The flagging economy, political and student opposition, prodding by international aid donors and the disintegration of the Soviet Union all had an impact on the viability of Ershad's regime, according to Guhathakurta. The 'politics of rebellion' were aimed not at overthrowing the government by violent means, but rather at forcing it to resign and transfer power to a caretaker government. However, with Ershad's resignation the movement that had brought about his downfall quickly came unstuck during the election campaign, leading to violent clashes.

In *Militarization, Violent State, Violent Society: Sri Lanka*, **Jayadeva Uyangoda** states bluntly that 'state-centric and counter-state violence represent a qualitative shift in the dominant mode of political bargaining in South Asian societies'. Part of that shift has been the breakdown of the social consensus that existed in Sri Lanka after Independence. Another factor has been the development of 'unformalized' agencies of state violence – officials who carry out extrajudicial killings and unofficial death squads – which 'can no longer be treated as secondary to the formal organs of the state such as the legislature, the executive and the judiciary. They are substantially the state.' With opponents of the state rejecting the use of democratic institutions in favour of violence, what in effect is happening is the negotiation of a new social contract by means of violence. However, the war between the Sri Lankan state and the Liberation Tigers of Tamil Eelam has been bred by ethnic essentialism, which breeds violence, which in turn reinforces essentialism and exclusivity. With no let-up in the spiral of violence and no space for interventions in the political debate in favour of non-violent solutions, the author is pessimistic about the future of democratic solutions in Sri Lanka.

Tanika Sarkar examines the growing visibility and militancy of women in Hindu nationalist movements, particularly within the Hindu right-wing party, Rashtriya Swayma Sevak Sangh and its mass fronts, the Vishwa Hindu Parishad and the Bharatiya Janata Party. *Hindu Women: Politicization Through Communalism* describes the mobilization of women from mainly conservative backgrounds, who 'are speaking their own minds, in their own words' and who have acquired 'a new and empowering self-image'.

Part of that empowerment involves lessons in yoga, sword and lathi play, judo, the use of stenguns, ideological discussions and discipline, all aimed at training for war against Muslims, as well as self-defence in a society rife with dowry murders and violence by husbands. The author notes that the Hindu women's movement has helped 'hitherto homebound women to reclaim public spaces, to acquire a public identity; it confers upon them a political role and even leadership'. But it has also cultivated involvement 'in a violent campaign of blind hatred' and prepares women 'for citizenship in an authoritarian Hindu state, and to reject secular, democratic politics'.

Pakistan in the 1980s was the scene of growing political activization of women at two levels: one, in response to the repressive measures depriving women of what were seen as their established rights, and the other in support of the ethnic-national cause. In *The Gender Dimension in Sindh's Ethnic Conflict*, **Khawar Mumtaz** explores the participation of women in the Sindhi nationalist movement and its impact on them. The author provides historical background on the development and organization of the Sindhi nationalist groups and the Mojahir Qaumi Mahaz, before detailing the role of women on both sides of the ethnic divide in Sindh. While finding the Sindhiani Tehrik and the MQM's women's groups different in their political and social goals, organization and degrees of openness to outsiders, Mumtaz also notes shared characteristics of class backgrounds in first-generation educated women, 'their energy, commitment and militancy, as evident in public demonstrations and rallies, their ability to handle weapons'. However, given the intensity of ethnic politics in Sindh, the author concludes that the extent of women's involvement in the ethnic-national movements and their priorities will continue to be determined by the predominantly male leadership of the movements.

In *Strategies for Conflict Resolution: the Case of South Asia*, **Kumar Rupesinghe** outlines several possible options for mitigating or resolving conflicts in the region, after describing general trends in the region, the role of the state, the evolving concept of sovereignty and a general typology of conflicts, including interstate conflicts, governance and authority conflicts, ideological conflicts, identity conflicts and resource-based conflicts.

In the context of South Asia, broad suggestions for transforming conflicts have included greater degrees of autonomy for minorities, fundamental social reforms, political democracy or consociational democracy. However, Rupesinghe stresses that a 'rationalist formula' may not be suitable for dealing with all the phases of conflict. In his native Sri Lanka, the author points to the potential of the Sangha as a mediator of anti-Tamil violence, but only if the Buddhist religious community 'can develop their own role in the peace process and their own approach to reconciliation'. Rupesinghe also notes the experience of peace groups in the Philippines which have been instrumental in the establishment of 'zones of peace', as well as peasant groups in Colombia, who developed methods of dialogue with guerrillas and government forces. In South Asia, he suggests, parties to conflict may be able to borrow some elements from the preventive and peace-maintaining experience of the European Union (internally), or the evolving capacities of the Organization for Security and Cooperation in Europe to develop instruments for dispute resolution, such as the OSCE's High Commissioner for National Minorities, or its Conflict Prevention Office.

1

Ethnic Conflict: Rethinking the Fundamentals

KUMAR DAVID

1.1 Introduction

The terms 'ethnic' and 'ethnicity' are used here to include religious, racial, linguistic, tribal and similar divides which have been activated in sociopolitical conflict in the present age. The use of a single generic term is justified by the palpable fact that the common features of these conflicts greatly overshadow the specificity of their religious, racial etc. character. A religious conflict in one area may have more in common with, say, a linguistic problem in another place, than with another religious conflict. The unfolding of events in a specific case depends much more on the particular political antecedents, economic conditions and problems of state, than on whether the phenomenon manifests itself as, for example, a language or a religious conflict. This chapter is basically concerned with ethnicity in relation to political conflict in the present period, and categories and concepts are advanced for this purpose.

The chapter takes it as an agreed and given fact that ethnic conflicts have assumed major proportions and have become an important feature of political life in the final quarter of this century. Further, that the scale, intensity and persistence of these conflicts provide persuasive evidence that we are not dealing with ephemeral events or accidental reflections of some other predicates. That is to say, it is taken as agreed that ethnic conflict cannot be reduced to a distorted or indirect reflection of class conflict – it is asserted that such reductionism is false. This is not to deny that class conflicts are themselves also fundamental determinants of history but rather to emphasize their intertwining with ethnic issues.

1.2 Ethnicity as Category

Superficially, ethnicity is associated with a sense of identity arising from shared customs, language and culture, physical characteristics, and so on, and would appear to be far removed from the material categories of economic and social life. In the context of this discussion, however, this is a highly inadequate appreciation of the role of identity as a political factor. True enough, ethnicity as a category in modern political conflict exists at the level of consciousness; but it is a reflection in consciousness of very real, concrete and material circumstances.[1] Consciousness does not reflect material reality in some mechanistic way; and indeed a consciousness of ethnic identity can persist long after the material foundations that engendered it have withered away, or it can emerge in advance of the proper consolidation of an immanent identity.[2] Nevertheless, there is a firm causal link between the consciousness of ethnic identity and the material organization of social life.

The material circumstance that underlies an ethnic unit is illustrated most simply by drawing attention to the periphery, or boundary of a socio-economic unit which possesses its own internal structures or mode of production. This is most obvious in the case of the separate tribes or the homogeneous kingdoms and nations of previous ages. A well-defined and specific set of people, a common territory (land, water, wildlife) and frontiers that must be guarded against invasion, delineate the boundary of production. The boundary, or periphery of a mode of production, is a material fact – a feature it shares with the productive powers of society and its internal or class divisions.[3] The sense of identity and the sense of security derive, at the level of consciousness, from the material reality of belonging within the mode of production, of being within the periphery.

A stark example is South Africa, where the Boer worker has identified with Boer landowner and bourgeois and not with his black class-brother. But these identities have been consolidated by separate economic existence, and its concomitant of war, conquest, slavery and territorial expansion, from the earliest settler times. In South Africa the Boer and the numerous African economic units were historically distinct enclaves, distinct modes of production. Only much later did modern capitalism supplant this, mainly after the consolidation of British Imperialism at the turn of the century. The extreme form of apartheid consciousness and of separate identity is deeply rooted in a now largely defunct, but at one time very real, subdivision of the economic universe into distinct and warring units and modes of production – war, subjugation and peonage being in those times but an adjunct to economics. This is evident in the following extracts:

> Their [Afrikaner or Boer] mode of production was simple. They could not build industries; their farming was subsistence; and they partook in only limited commerce. To build their subsistence economy, they had to depend on slave labour, and in this spirit they killed, dispossessed, those they found occupying the land they coveted. They found their identity in the negation of those they conquered and exploited . . .

> When the Boers were forced to migrate to the interior by the threat of British occupation in the 1820s, they did so in order to maintain, not to improve, their economy and standard of 'civilization' . . . 'fundamental innovations in the use of land or in social practice were not made in their minds' . . .

> It is in the light of this process that racial nationalism must be understood. The Boers' espousal of a doctrine of African inferiority, justified on biblical grounds, was interconnected with their desire to justify peonage. And why not? These people had

inherited from their settler forefathers feudal-like institutions with rigid hierarchical structures. For their ancestors, 'race' had provided a suitable principle on which to create a servile population. Their religious leaders found in the Bible the 'Curse of Canaan', which they adapted to justify their activities. (Magubane, 1979, pp. 32–33)

The theoretical approach developed here also provides the answer to the question of why some ethnicities are activated but not others. For example, why do Tamils, including Tamil Christians, take one side and Sinhalese, including Christians, the other? Why not Buddhist against Christian, instead of Sinhalese against Tamil? The answer lies in seeking out which material identities were historically separate and significant, and hence became consolidated into ethnicities; which material boundaries of cohesive socio-economic units formed important peripheries to modes of production. Tamil Christians were internal to and a part of Tamil (or Jaffna) society, economy, trade, agriculture and so on. Similarly, Sinhalese Christians were internal to another cohesive socio-economic unit. In Sri Lanka, there is no Christian or Buddhist ethnicity: there are Sinhalese and Tamil ones. In the Punjab, or in Northern Ireland, the superficial division is precisely the opposite, religion being the determinant of cleavage in the ethnic civil wars. The argument about historically sustained economic separateness functioning as an original delineator of ethnic identity, however, shows no such contradiction as superficial theorization does, and remains wholly plausible in all of these examples – notwithstanding many oversimplifications in conventional Left thinking, for example that of Bipan Chandra.[4]

This approach also explains why, however slowly and tortuously, ethnicities disappear or become politically irrelevant. What became of the Vandals and the Goths? How did German ethnicity replace these groupings? Whatever happened to the Normans and the Saxons – and who on earth is an Englishman? Why are the Protestants and Catholics in England, or in Germany, not at each other's throats, in emulation of their brothers in faith in Belfast? The answer lies wholly in material categories – membership of a common mode of production, indifferentiability of material intercourse and the consanguinity which follows from this were realized in Germany and England a long time ago. Such too will be the future of the USA, that great melting pot of innumerable peoples, where ethnic particularity appears destined to disappear.

Though ethnicity as a category in political conflict and the concomitant consciousness of identity are linked in this way to material life, it would be a mistake to overlook their dis-synchronous time scales of development and the consequent relative autonomies of their particular dynamics. For example, it would be hard to argue that the Boers and the blacks of South Africa do not even today co-habit the same material socio-economic entity. But it is obvious that consciousness of racial identity is not changing at a pace synchronized with these material transformations. Ideology will finally align itself with reality, but not until it has exhausted its own tortuous and extended life-span. Furthermore, the very history of material change has brought about a new overlapping of class with race in South Africa (Magubane, 1979, pp. 163–192), providing a new and more complex rationale to perceptions of ethnic division and identity.

1.3   New Nation-States

In current discussions of ethnic conflict it is not uncommon to hear remarks about the 'arbitrary boundaries of new artificial nation-states which have been created by colonialism'. What is partly at least implicit in this is the assertion that these nations are entities which are in some way irrational and have little right to exist. What is at issue here is not colonial conquest per se, but rather one specific aspect of the colonial legacy: the new nation-states as they actually exist. It is necessary to re-examine the immanent critique of the right of such 'artificial' entities to exist. But the very posing of the question raises a difficulty – surely it is not feasible to suggest that the world should return to the patchwork mosaic of myriad kingdoms, tribes, principalities, states and fiefdoms, each in the modern sense a separate state or semi-state, that preceded colonial conquest? For that matter, surely it is not possible to suggest that Germany should return to the dozens of independent or semi-independent units that preceded the conquest and unification of these lands by Napoleon?

It is worth pursuing this argument a little further and pushing to its limits the view that India, for example, is an artificial entity created by British Imperialism. Implicit in some formulations of this assertion could lie the value-judgement that the subcontinent should be divided into six or 16, or whatever, ethnically more homogeneous entities. The examples can be multiplied. Why not Matebeleland and Shonaland instead of a single Zimbabwe, why not a separate Quebec and a separate Tamil Eelam and a separate Moro-Philippine, and so the list goes on. We must take this argument at face value and respond to it seriously.

Looking at the world as it has emerged out of centuries of colonialism and its legacy of forced amalgamation of peoples and races, and taking into account the numerous separatist movements that exist today, it is perfectly reasonable to take the following hypothesis as a serious agenda for discussion: 'What is so magic about the number 150 (or so), what is wrong with a world consisting of 300, or for that matter 450, nation-states? Long live the slogan: A STATE FOR EACH ETHNICITY!' The hypothesis, of course, fails at first sight because it can simply be seen to be absurd, but this does not amount to a considered refutation. A considered response, formulated in a general way, as to why the subdivision of the world into an ever larger number of smaller entities, or, to put it more starkly, why the subdivision of India into say 16 ethnic states, would be irrational and reactionary, consists of the following three points:

- Larger national units, without carrying the idea to absurd limits, and taking into account specific and practical constraints in each case, are capable of marshalling their resources more efficiently and organizing their economic production more rationally – for example the USA and China.

- A mosaic of small nations all with their own customs barriers, passports, national anthems and flags would be a definite hindrance to the full development of the spiritual, intellectual and cultural potentials of the human species. The removal of artificial impediments of this nature within the European Union, for example, quite apart from its intrinsic economic rationality, and the way in which people are taking advantage of this enhanced freedom of movement and interaction, are pointers to how a more unified world of the future should look.

- Integration into larger national states in the case of the less developed countries reduces the scope for external exploitation by more powerful imperial or neighbouring states. For example, a separate Puerto Rico on the borders of the USA, and Puerto Rico as the 51st state in the federation, are very different entities in this respect. Or again, China is not Vietnam in relation to Soviet interests.

There are very strong economic and spiritual-intellectual-cultural reasons for dispensing with divisions and integrating people into nations which owe their raison d'etre to considerations other than ethnic particularism. To complete the discussion, however, I must point out that against these arguments must be considered the possibility of economic exploitation of specific ethnicities within an expanded state, the possibility that a minority identity or culture is oppressed within an existing state, and the likelihood that a heritage of backwardness may make the largeness of a unit not an advantage but a problem in relation to efficient and rational economic management. The generalization that has been previously attempted therefore is not without exceptions; but the real significance of the generalization lies not in the fact that the exceptions are less numerous than the norm, but rather that for these exceptions too the generalization points to the long-term future.

Let me mention briefly a few examples which to my mind are, on balance, exceptions to the generalization for one or another reason. Clearly the dismemberment of the old Pakistan into Bangladesh and Pakistan was a step forward; probably it is better for all concerned that the Baltic states have ceded from the Soviet Union; it is difficult to justify China's continued occupation of Tibet. I make this last remark without prejudice to the arguments put forward by each side about whether Tibet is, or is not, historically a part of China.

These are indeed possible exceptions, and in specific cases, indeed, the particular history and circumstances may be compelling. The concept of an exception, therefore, needs to be located and understood more precisely. There is a historically progressive and general trend towards the integration of nations politically and economically, as well as culturally, into larger entities. However, while this trend has been self-evident over time-spans such as centuries or decades, the process is also an uneven one, mediated by concrete and specific factors of shorter historical duration but of great, though ephemeral, intensity. Hence, reversals of the general trend will take place, from time to time, in specific instances and along one or another of its axes (that is, political (state), economic and/or cultural axes). It would be correct in certain such instances to support some specific 'reversal' if in the final analysis it is conducive to the longer progressive historical trend. However, even in doing so – that is, in formulating the particular form and nature of the support to be extended – the longer-term progressive world historical trend previously discussed must be constantly kept in mind, and in the long run must assert its priority. This is the essence of the concept of exceptionalism.

It appears that those who say that modern India is a creation of the British Raj have forgotten to add that it is nevertheless an irreversible one. The meaning of the concept of irreversibility needs to be stated more precisely, since its main thrust is an intention not to be prognostic but to be provocative. Economic production and the market in India have been sufficiently well integrated that India's continued existence as a unified nation is in the interests of all of the classes of modern society, the bourgeois and proletariat included; and the strength of these classes is likely to overcome fissiparous pressures from remnant classes of previous (pre-modern) society when they do arise from time to time, including even those instances when ethno-specific sections of modern classes may participate in such a divisive alliance. Secondly, modern cultural contents are generally strong enough to overcome fissiparous remnants from previous historical times. This is the essential content of the concept of irreversibility as used here; it does not purport to make prognostic statements about what might happen in, say, Kashmir, but rather it is a concept whose meaningfulness arises in the context of the general thesis of human progress that underlies this discussion.

The demise of the Soviet Union gives rise to some important attitudinal questions. The unresolved confrontation regarding the nature of the economic system in the nations issuing from the former USSR is shot through by a perpendicular emergence of ethno-politics and widespread armed nationalist conflicts. We are witnessing, simultaneously, a political revolution (the overthrow of Stalinism), a social counter-revolution (the attempted restoration of capitalism) and also a resurgence of ethno-politics. The class–state axis, that is to say the question of the restoration of capitalism, remains the supreme issue of the moment. As the resolution of this issue mediated by the intervention of world capitalism works itself out, and irrespective of whichever direction this resolution takes, a period of nationalist political conflicts and wars will follow. It is not entirely unreasonable to suppose that the disintegration of the Soviet Union may produce some economically viable and politically stable nations; nevertheless it remains paradoxically true that ethnic conflicts are the greatest impediment to long-term progress.

The importance of understanding each particular example on its own terms places the methodological emphasis on concrete analysis, with abstraction and generalization forming a background of knowledge. Or as Clive Thomas (1984, p. xx) says:

There are . . . further methodological advantages to this approach. One is that the use of concrete examples allows us to study simultaneously the similarities and differences in the form that the state takes in peripheral capitalist societies. We can then avoid two dangers. One is oversimplification – that is, an approach that is premised on the view that since each society is different, that there can be no general theories; the other is overgeneralization, which results from a preoccupation with similarities.

Thomas was studying authoritarian states in peripheral capitalist societies. His methodological remarks, however, have general relevance to other issues as well, such as countries with ethnic unrest, be they capitalist, peripheral capitalist, Stalinist or post-Stalinist – which topic constitutes the brief of this chapter.

1.4   State–Nation–Class

Let us begin by examining the nature of the state in newly independent countries and its interaction with class formation and the national question. The first point that needs to be made is that as colonialism withdraws it does not leave behind a society with a strong potential ruling class in place. The other side of this same coin is the economic backwardness and the weakness and distortion of the productive forces in these countries. From the beginning, therefore, the state is an unstable and tottering structure. The most primitive form is the military state, governed by the crude violence of a body of armed oppressors in a manner reminiscent of the ordering of the proto-state among old barbarian hordes. The junta, in the case of the smallest or most unstable military regimes, rests on the narrowest of possible social bases, the military itself, which is held together by the ability to plunder (directly or indirectly through government), and survives by sheer violence until overthrown by another armed horde similarly intent on plunder. Even if less transparent in some cases, this is the taxonomical genesis and quintessential character of all Third World military dictatorships.

In larger countries with military regimes – Pakistan, Burma or Thailand, for example, in Asia, or numerous examples in Latin America – the formal class structures are better formed. Although the ruling class is not always able to sustain power at all times on its own within and through civil society, it does have its residues of strength. It also has its liberal and democratic segments who despise the uncouthness of the gangsters in khaki uniform. The relationship between the military regime and the class basis of government and state is now a more complex and changeable one. In this unstable environment, with both the ruling class and the sections of the military leaning for support on narrow and specific social segments, ethnicity is frequently activated as a political dimension. This can happen in various ways: for example, the armed bands themselves may be racial, tribal or religious hordes, or the junta's survival may be based on sectional appeal. Hence, military or military–civilian musical-chairs regimes in multi-ethnic societies, in the last few decades and in the present period, are likely to be divisive. In ethnically homogeneous societies they tend to be rapacious instruments of class domination and generalized repression; in multi-ethnic societies, divisive instruments of particularized repression and incipient civil war.

Not all the tottering postcolonial states have military regimes: many have long traditions of civilian rule – Colombia, India, Kenya, Mexico, the Philippines, Sri Lanka, Zambia, Zimbabwe and so on. In several of these places ethnic conflict is active or acute. The underlying developments are not different from the previous examples, although their manifestation in the political arena appears to follow a different course. Sri Lanka is perhaps the classic example, in the sense that the dynamics are particularly transparent and could be used as a laboratory model for exemplary case-study. The motivating causes of a class alliance between the ruling classes and an emerging Sinhalese petty bourgeoisie, the inevitability of ethnic polarization as a consequence, the emergence of a hegemonic chauvinist ideology, the transformation of the state into an instrument of chauvinism and the growth of Tamil nationalism and terrorism, in part as a response to state terrorism – these have been described at length elsewhere (David & Kadirgamar, 1989; Kadirgamar, 1989; David, 1990). Similar dynamics often occur in other places; although the connection between basic class and state 'mechanics' and manifest political events may not be so transparently clear, the linkage between state, class and ethnic conflict is always fundamental.

Some important generalizations of theoretical significance for this section are summarized briefly below.

- Weak ruling classes are driven into unholy political alliances, usually to overcome social instability arising from mass poverty and/or a radical political challenge.
- When an alliance of this nature involves the petty bourgeoisie of some/one of the particular ethnic groups in a multi-ethnic nation, the fundamental fact that *petit bourgeois* ideology is inherently and inevitably divisive stamps itself on the nature of the class alliance and on the ideology of the state.
- Such states become instruments of ethnic repression, and conditions are soon rife for civil war. Paradoxically, however, this state is simultaneously, and in a sense involuntarily, also playing the historically progressive role of preserving larger, and more rational, state entities.
- Numerous international factors and internal variables specific to each country contribute significantly to the particular dynamics in each case.

The paradox noted is at the root of the great degree of autonomy that the state enjoys in these countries. Clive Thomas (1984, pp. 80–81) describes it as follows:

> In many peripheral capitalist societies the level of internecine struggles is so high that there is a real possibility that the society will disintegrate. Further, because no class is strong enough to impose its solution to the society's problems, a class outlook

does not dominate political relations. But even if it did, it must be remembered that class struggle creates the opportunity for revolutionary advance but does not automatically guarantee this advance. Every class-divided society is faced with the risk that class struggle may lead to its disintegration and to the 'common ruin' of all classes. And, if we add to class struggle struggles of a racial, religious or ethnic character, we can see that the risk of 'common ruin' only increases. In such circumstances, the state's claim to preserve the integrity of the society is appealing to all classes, and this in turn favors tendencies that support the autonomy of the state.

    . . . it must be emphasized that the autonomy of the state cannot be viewed one-sidedly, that is, as a reflection of the status of contending classes and other social groups. This autonomization is itself a necessary factor in ensuring the consolidation of the dominant mode of capital accumulation. In addition, in so far as state investments and state controls are a part of this consolidation and are linked to economic backwardness and the underdevelopment of the major classes, this autonomy *is not necessarily of a short duration*. On the contrary, the 'exceptional circumstances' that favor state autonomy in post-colonial societies may last for a long time indeed.

Before resuming my main theme, it is worth noting that the remarks about the sustained autonomy of the state and the consolidation of the dominant (not necessarily capitalist) mode of capital accumulation in backward societies are of interest to the student of the longevity of Stalinism as well. The challenge to the alienated state and/or to the mode of production also in Eastern Europe and the Soviet Union is a complex one – will the socialist baby be thrown out with the Stalinist bath water?

An examination of Thomas's discussion, which begins with class, permits several important adjustments and insights when extended to the discussion of ethnicity. The first point of importance is that 'a class outlook does not dominate'. Indeed, where ethno-politics has reached the dimensions of critical confrontation, the state–ethnicity interaction 'dominates'. This is better described by the use of Althusser's category of overdetermination.[5] In a multiethnic state where no class or group is strong enough to impose 'its solution', the overdetermining axis of social existence and conflict is the state–ethnicity axis. The state becomes an instrument of ethno-politics; every aspect of social – and for that matter private – life is driven by ethnic considerations: jobs, education, national development and investment, constitution, government and army. Conversely, as the dialectic of overdetermination implies, each one of these aspects of social and private life, in turn, also becomes an arena for the aggravation of ethnic friction.

The second significant point to reconsider is the scope for the realization of the relative autonomy of the state. As Thomas rightly points out, faced with the 'common ruin' of the contending classes, the state rises above class conflict as a (false) saviour and consolidates its arbitrary powers. However, where society is overdetermined by ethnic and not class conflict, there is a significant difference. In ethnic conflict, specifically when separatism is the central question, *the legitimacy of the state itself is the issue*. Hence, it may happen that what becomes a possibility is not the autonomy, but its very opposite: the rejection and indeed the physical elimination of the state and its substitution by another – Jaffna, Eritrea or the Moro regions of the Philippines.

Consequently, because belief and trust in the state, and the proto-state (one on each side of the ethnic divide) are greatly tarnished in the minds of all classes and 'races', the state and the proto-state both become increasingly repressive. This is the only way these entities can survive, once bereft of their moral basis. At best, the state and the proto-state may enjoy short-lived periods of ideological hegemony during moments of military victory over the ethnic 'enemy'; but true political hegemony and moral acceptability among their own peoples will always evade them. Physical suffering, in those cases where conditions of war prevail, further accentuates this alienation of the state and the proto-state from their respective ethnic populations.

Ethnic instability is not a peculiarity of the Third World. At the moment of writing Yugoslavia is tearing itself apart like a house possessed; the future relationship of the republics of the former Soviet Union is a matter of much contention; and there are Ulster, Quebec and the Basque provinces, to mention but a few. Yugoslavia appears to be an appropriate reference point for a few relevant remarks about the problems of consolidation of nation-states.

The consolidation of several South Slav nationalities (tribes and kingdoms) into a single state took place in tortuous ways that can be traced from the fall of the Austro-Hungarian Empire up to its realization in Tito's Socialist Yugoslavia. Throughout this period, and up to now, ethnocentrism, Serbian domination, Croatian extremism, Albanian 'irredentism', the amputation and partitioning of Macedonia between Yugoslavia, Bulgaria and Greece, and such-like issues, have remained now dormant, now threatening, on the political landscape of the modern nation-state. The progress that has been made, though halting and interrupted, has to do with two imperatives: a recognition by South Slav nationalists of the mutual advantages of Slav unity, and the natural tendencies that flowed from the ideology of Yugoslav Communism. To quote Z.T. Irwin (1984, pp. 107, 109, 119):

In 1918, the idea of a slavic union appeared to be a logically completing consequence of the collapse of the Austro-Hungarian Empire, Serbia's role in the war, and the principles of self-determination. It was assumed that Croatia, Slovenia, and Bosnia-Hercegovina could best insure their independence in union with independent Serbia. The Yugoslav idea, literally the union of South Slavs, enjoyed prestigious intellectual spokesmen . . .

In 1939, the Serbian Premier, Dragisa Ćvetković, and the Croatian leader, Vladimir Maček, signed a *Sporazum* (understanding) which recognized Yugoslavia as the 'best guarantee of the independence of the Serbs, Croats and Slovenes.' The *Sporazum* promised autonomy for the Yugoslav nation by the creation of different *banovi* (provinces) and a local Croatian assembly . . .

In his address to the Eighth Congress of the (Communist) League, held in 1964, Tito attempted to criticize equally the 'bureaucratic-centrist tendencies' and 'unitarianism [that] ignored the socio-economic functions of the republics and the autonomous regions', as well as the 'tendency to be shut up within one's "own borders"' . . .

Yugoslavia's communist leadership had attempted to steer a course between policies dependent on excessive central power and those which would lead to local fragmentation.

As new nations struggle towards the consolidation of nation-states, two crucial lessons can be learnt. First, periods of consolidation will alternate with periods of crisis and rupture; and secondly, later periods may include the re-drawing of the boundaries of these nation-states. Furthermore, the time-scale inherent in nation formation is an extremely long one and is deeply punctuated by economic success and failure. In economically advanced countries like the USA, Canada, Italy, Switzerland, Germany, France, the United Kingdom, and so on, although ethnic conflict is far from absent, the issue of redefining national boundaries is not of real significance.

The general hypothesis that this analysis leads to is that while class, state and ethnic variables (or political and military factors, to state it in another way) account for the sharp turns and sometimes irreversible ruptures and rearrangements, in the long run the consolidation of the nation-state will be determined mainly by the economic success of the prevailing mode of production. To put it crudely, there are two dis-synchronous time cycles at work, and they impinge on each other and are partly determined by each other. We have to think of *overdetermination* as a dynamic concept describing a changing reality, and we must understand that the impinging of these two different structures on each other mediates the metabolism of change. The consolidation of new nation-states is imbued with an uncertainty rooted in this dis-synchrony of the determining events and variables.

1.5   The Limits of Conventional Solutions

The turmoil arising from ethnic conflict has been with us for several decades now, and many 'solutions' have been described, and attempted. To the extent that all these answers have been around and/or been attempted in practice for this long, they are no longer new or radical, so let us refer to them as conventional solutions. To this category belong 'solutions' of the Left and the Right, of oppressors and the oppressed, and include the following:

- Forcible (mainly military) integration, incorporation or elimination of recalcitrant ethnicities.
- Civil wars, national liberation struggles, separatism.
- Federalism, regionalism, autonomy, devolution, democratization, economic decentralization etc., implemented to various degrees and in various forms.
- Statesmanship and its opposite political chicanery, linkage between class politics and self-determination concepts, ethno-coalition politics.
- Foreign interference, intervention or intercession, by other countries and/or various agencies such as the UN, the EU, the IMF, human rights and peace movements.

Many or all of these factors will be at work, with one or the other the central strategy at any given time. Here we shall explore the limits of such solutions in general terms; although this analysis is based on the experiences of the past four decades, explicit references to individual cases will be avoided for the sake of brevity.

The first assertion is that today, unlike in previous centuries (the colonial and the settler periods), a forcible or military solution is exceptional, if not impossible. That is to say, a military solution to a mature ethnic movement, whether by its defeat or conversely by the victory of separatism, is very exceptional indeed. The reasons lie in both the changed nature and balance of world politics as well as in world technological changes and the near-universal accessibility of this technology, albeit at a price.

The second important feature to understand involves the complex, and in a sense peculiar, ways in which constitutional re-arrangements and enhanced democratization can effect ethnic instabilities. Thus, for example, the transition from a repressive regime under which ethnic tension lay invisible, to a more democratic one which sets about attempting to restore greater autonomy, may lead not to a period of compromise and harmony, but rather to a period in which various extremist tendencies gain ground, narrow chauvinist ideas triumph and ethnic clashes in society multiply. The root cause here has to do with the fundamental limitedness of ethnic consciousness itself, to be discussed later in this section.

Much has been said about democracy and autonomy/devolution being the cornerstones of a solution to ethnic conflict. Great faith has been placed on this approach by democratic peoples' movements within affected countries and by international human rights and peace agencies. Undoubtedly, as a set of core ideas these assertions are valuable; the point, however, is that their limitedness as a complete programme has not been sufficiently drawn out and discussed. The limitedness arises from two sources. The first is the limitedness of the ethnic consciousness from which democratic and devolutionist solutions, if applied in isolation, cannot separate themselves. The second is that economic devolution, to the extent that it is included in an ethnic 'settlement', leads not to a solution but only to a subsequent conflict 'at the boundary'; and to the extent that it is not included, makes purely political devolution a mere farce. There is no way out along the purely 'liberal-democratic' road. The failures of such efforts in the recent decades illustrate this, since it is not convincing to assume that, say, political chicanery can explain why not even a single reasonable counter-example is available.

There is thus a fundamental contradiction between ethnicity as the embodiment of the identity of a separate consciousness (arising from and carrying the stamp of an isolated mode of production), and the reality of modern nation-states (and indeed the modern world), where the integration of the mode of production is far advanced, and material intercourse is universalized between different peoples and inextricably intertwined between nations.

At this point, some discussion of ethnic consciousness is in order. There are several conflicting value judgements that have to be ordered and reconciled. There is ethnicity as the specificity, the richness and the repository of the culture of a particular people; there is tolerance and respect for all ethnicities and the call for a 'celebration of plurality'; there is identity as a haven of security and hope for oppressed and exploited races and religions; there is ethnicity as a narrow identity in a material world which has far outgrown the origins of the consciousness of separate identities; there is ethnicity as a politically and morally divisive influence; there is ethnicity as racism, chauvinism and prejudice. Apart from the trite observation that ethnicity is good but too much of it is bad for the body (politic), liberal philosophy has not had much else to say. Something more is crying out to be expressed.

We must recognize that ethnic consciousness, in the final analysis, is a remembrance of things past, and as mankind grows it will, in the words of St Paul, 'put aside childish things'. Surely, there will be a universalization of our heritage instead of an eternal particularization of it? The sense of identity and security that particularity provides, and which indeed is so important at times today, must nevertheless be seen as an ephemeral phase in the longer journey that mankind has undertaken. When human beings circumnavigate the sun and settle on Mars will they still carry their ethnic identities with them? Perhaps – and this gives rise to the need for some remarks regarding ethnic ideology of a more base kind.

I am using the term 'ethnic ideology' as differentiated from 'ethnic consciousness' to denote the base elements of racism, intolerance, prejudice and chauvinism which are a part of the ethno-political scene. Such ideologies are still deep and divisive all over the world; they are not confined to small numbers of less enlightened individuals or to extremist organizations.

> The full extent of the deep, and subtle, and often unconscious penetration of the consciousness of most Indians by aspects of communal ideology is seldom realised by most of us. (Chandra, 1984, p. 10)

The ever-so-comforting assertion that, say, racism or communalism does not run deep in the ordinary people, who are but innocents misled by guileful politicians, is a naive oversimplification. Divisive ethnic consciousness, chauvinism, racism and religious intolerance are ubiquitous ideologies that may run deep among the people in various ethnic groups. It flies in the face of empirical evidence to assert during an epoch of sustained ethnic conflict that the rural folk, the ordinary man and woman, the middle classes, the worker, and so on, are free from prejudice like the noble savage and are simply the victims of false leaders and opportunist politicians. Ethnic ideology has a deep grip on mass audiences, and false prophets and opportunist politicians may be more a result than a cause. We can borrow this quote from Thomas, about the authoritarian state, and read it quite meaningfully with ethnicity in mind:

> ... despite the notoriety of the Shah, Bokassa, Somoza and Gairy, and despite the unmistakeable influence they have had on the state and on political forms in their societies, it is not the leaders who determined the character of these states – they are more effect than cause. Consequently, as we shall see, the authoritarian state cannot be reduced to the existence of a dictator or to authoritarian and dictatorial

forms of rule, although these accompany it. We must look at the state as a historical materialist category and understand its social and material basis. (1984, p. *xx*)

If, for example, Sinhala chauvinism is a fact, it is then also a deep reality of
the consciousness of the corresponding people. If one is to move forward,
then, the long fight against false ideology is a major task that cannot be
avoided. This is a sustained struggle, and for long years the task will fall on a
few with the vision and the courage to bear it. A whole epoch of disappoint-
ment and defeat will precede tangible achievements in the larger social arena.
Progress will be slow and difficult because ethnic ideology has old and deep
roots which have been reinforced by modern social and political conflict and
economic crisis. The great mistake, however, is to fail to realize that a new
rational and more civilized world cannot be born unless the ideological and
philosophical struggle to free human minds from the limitations of ethnicity
is undertaken and an adequate commitment made. Many well-meaning
organizations have not managed to grasp this nettle.

1.6   About Dialectics

It has been argued above that at a sufficiently fundamental philosophical
level, the ideology of ethnicity must be rejected as false consciousness; that
the economic unification of the world is irreversible. It is also true that modern science implies the universalization of knowledge. As barriers break
down, cultural intermingling proceeds apace. Simultaneously, it has been
argued that the rights of oppressed nations to self-determination must be
upheld; democratic and cultural-linguistic-religious rights of ethnic minorities
must be protected; and that a plural, and by implication secular, society must
be advanced. Do these two assertions contradict each other? I think not. It is
not a contradiction to accept the unavoidable limitations of the world as we
find it while undertaking at one and the same time a commitment to ending
such limitations. Surely, it is not contradictory to say, as did Marx (1844),

> Religion is at once the expression of real oppression and the protest against that
> oppression. Religion is the sigh of the oppressed creature, the heart of a heartless
> world, the soul of soulless conditions . . . To abolish religion as the illusory happi-
> ness of the people is a demand for their real happiness. The call to abandon illusions
> about their conditions is the call to abandon a condition which requires illusions.
> Thus, the critique of religion is in embryo the critique of the vale of tears whose
> halo is religion . . . Criticism has plucked the imaginary flowers from the chain, not
> so that man may bear the chain without fantasy or consolation, but so that he can
> cast off the chain and gather the living flower . . . so that he shall think, act, and
> fashion his reality as a man who has lost his illusions and regained his reason, so
> that he will revolve about himself as his own true sun.

and at the same time to demand the freedom of religion for all men!
    Another reason for quoting this well-known passage at some length: there
is a parallel between this criticism of the sociopsychological role of religion

and the criticism of ethnic ideology as false consciousness in this chapter. The point then is that a true approach to ethno-politics must run simultaneously at two levels. If it is not only a commitment to oppose racial or religious oppression, to stand up against oppressive regimes and social orders, but also in the final analysis the commitment to a new world order and a new universal human consciousness, then the issue of class and social justice is inseparable from the issue of national and ethnic justice. No organization which fails to link these two aspects of social change within itself can achieve lasting solutions to either.

1.7   Conclusion

Here I have argued that ethnic conflict as a modern political phenomenon is not confined to backward societies in which the state is still in the process of formation and consolidation, and that it will persist for a further period of human civilization. The events of the past few years and more importantly their underlying causes – sustained ethnic oppression and conflict based in part on the persistence of ethnic consciousness in civil society, which have festered for many decades prior to the recent explosive manifestation, and which manifestations are in any case only signposts of the ebb and flow of more fundamental trends – have amply justified this thesis.

I have also insisted that the activation, the catalysation, of some ethnicity, somersaulting it from a latent state into the sphere of real and intense political activity, can be understood only on the basis of a concrete, historical materialist, examination. It cannot be understood from an idealist analysis, in terms of a thesis primarily based on the 'philosophy', language, ethnic traits, ancient history, some supposed natural characteristics or consciousness, and so on, of a particular race, religion or people. This too has been borne out by recent events, which have furthermore dramatically justified the assertion that whether a problem is religious in one location, linguistic in a second, and racial in a third, is far less important than the specific socio-economic dynamics which actually drives the events forward. Theoretically, this has justified the introduction and use of the generic category 'ethnic' as a valid concept in the construction of modern political theory.

The dichotomous nature of modern ethno-politics has also been discussed, being at one and the same time an expression of a people's desire for liberation and a recrudescence of enmity and xenophobia. The concepts of overdetermination and time dis-synchrony are useful in thinking through the uneven and dynamic nature of the complex interactions between the different elements (economy, class, state, ethnicity) of a social formation. Certain reductionist approaches have been disputed and I have sought here to debunk *naïveté* or the underestimating of the depth of ethnic prejudice in the populace at large. What we need is a dialectical approach which attempts to reconcile what is feasible at a given time, with commitment to a long-term vision.

Notes

1. There is a wealth of literature attempting to define and discuss ethnicity in the modern world. Anthony D. Smith places ethnicity in the context of modernization (1981, 1983). Cynthia Enloe examines the issue in the context of power and authority (1973), and in her other writings. John Saul looks at the issues of class and tribe in relation to imperialism and the modes of production debate (1979, pp. 347–372). An essentially idealist discussion will be found in Walker Connor (1972). Readers are no doubt also familiar with a mechanistic incorporation of ethnicity into class analysis which is/was common at one time and is now dismissed as crude or mechanistic Marxism.

2. The emergence of a new ethnicity in the Jharkhand region is described by Javeed Alem (1989).

3. The concept of 'conditions of production' to supplement those of 'forces and relations of production' was first introduced by B. Borochov (n.d. pp. 157–165). A more recent discussion can be found in R. Munck (1986).

4. The following is taken from Bipan Chandra (1984, p. 13):

> Not only did Hindus or Muslims or Sikhs or Christians not form a nation or a nationality, they did not even form a distinct and homogeneous 'community' except for religious purposes. That is, they did not separately form a 'monolithic social structure' or a cohesive unit on a religious basis with common economic, political, social and cultural interests, or bonds or outlook. The religious coordinates did not coincide with the class, ethnic, linguistic, or cultural coordinates. There was no sharply etched or articulated interests of Hindus or Muslims 'standing in juxtaposition to one another'; in particular, the conditions of Hindu and Muslim workers and peasants was the same.

Historically, this assertion is not valid. The history of the Muslim invasions of the Indo-Gangetic plain followed by several centuries of war, the establishment of Muslim rulers, the attendant court and civil society around it, subsequent religious conversions and the linkage of conversions to castes which were oppressed by prevailing Hindu society, and finally the partial separation of Hindu and Muslim villages even up to recent times – all this is the history of societies which were to a considerable degree materially divided. These centuries constitute the historical roots of ethnic consciousness. Chandra's remark about the common interests of workers and peasants, in a current political context, is certainly true, but that is a separate dimension of the discussion.

5. The Glossary of Louis Althusser's *For Marx* (1977) provides the following explanation, prepared by the translator Ben Brewester:

> OVERDETERMINATION: Freud used this term to describe (among other things) the representation of the dream-thoughts in images privileged by their condensation of a number of thoughts in a single image, or by the transference of psychic energy from a particularly potent thought to apparently trivial images. Althusser uses the same term to describe the effects of the contradictions in each practice constituting the social formation, on the social formation as a whole, and hence back on each practice and each contradiction, defining the pattern of dominance and subordination, antagonism and non-antagonism of the contradictions in the structure in dominance at any given historical moment. More precisely, the overdetermination of a contradiction is the reflection in it of its conditions of existence within the complex whole, that is, of the other contradictions in the complex whole, in other words its uneven development.

References

Alem, Javeed, 1989. 'India: Nationality Formation under Retarded Capitalism', pp. 45–69 in Kumar David & Santasilan Kadirgamar, eds, *Ethnicity: Identity, Conflict, Crisis*.

Althusser, Louis, 1977. *For Marx*. London: New Left Review Editions.

Borochov, B., n.d. *Nationalism and Class Struggle: a Marxist Approach to the Jewish Question*. Westport, CT: Greenwood Press.

Chandra, Bipan, 1984. *Communalism in Modern India*. New Delhi: Vani Educational Books.

Connor, Walker, 1972. 'Nation Building or Nation Destroying', *World Politics*, vol. 24, no. 2, pp. 319-355.

David, Kumar, 1990. 'Sri Lanka: Is there a Way out?', *Capital and Class*, no. 40, Spring, pp. 15–24.

David, Kumar & Santasilan Kadrigamar, eds, 1989. *Ethnicity: Identity, Conflict, Crisis*. Hong Kong: Arena Press.

Enloe, Cynthia H., 1972. *Ethnic Conflict and Political Development*. Boston: Little Brown.

Irwin, Zachary T., 1984. 'Yugoslavia and Ethnonationalists', pp. 105–149 in Fredric L. Shields, ed. *Ethnic Separation and World Politics*. Lanham, NY: University Press of America.

Kadirgamar, Santasilan, 1989. 'Lanker: Nationalism, Self-Determination and Conflict', pp. 213–247 in Kumar David & Santasilan Kadirgamar, eds, *Ethnicity: Identity, Conflict, Crisis*.

Magubane, Bernard M., 1979. *The Political Economy of Race and Class in South Africa*. New York: Monthly Review Press.

Marx, Karl, 1844. Introduction to *Contribution to the Critique of Hegel's 'Philosophy of Right'*.

Munck, R., 1986. *The Difficult Dialogue: Marxism and Nationalism*. London: Zed Books.

Saul, John, 1979. 'The Dialectics of Class and Tribe', *Race and Class*, vol. 20, no. 4. pp. 347–372.

Smith, Anthony D., 1981. *The Ethnic Revival in the Modern World*. Cambridge: Cambridge University Press.

Smith, Anthony D., 1983. *State and Nation in the Third World: the Western State and African Nationalism*. Hemel Hempstead: Harvester Wheatsheaf.

Thomas, Clive Y., 1984. *The Rise of the Authoritarian State in Peripheral Societies*. New York: Monthly Review Press.

2

India: From Civilization to Nations

DEV NATHAN

2.1 Introduction

Discussions on the national question in India almost invariably assume that there is an Indian nation, something that has been handed down with a long history. For some, this Indian nation dates back even to the times of Mohenjodaro and Harappa. For others, it begins with the spread of settled agriculture and its attendant culture from the Ganga-Yamuna region.

There are a few who hold that India is multi-national, perhaps a multi-national 'country'. A problem with this second analysis remains to explain the notion of 'country' as distinct from 'nation'. In practical – i.e. political – terms, 'country' turns out to be no different from 'nation'. Further, there is the question: what of the undoubted cultural unity of India and, for that matter, South Asia? Is it as real and definite as the cultural unity of, say, Europe? Does the 'Indian nation' extend over all of South Asia?

For the Indian state, India's cultural or civilizational unity is the same thing as its national unity and national existence. In its *White Paper on the Punjab Agitation* (1984, p. 17) the Government of India stated:

> The Indian people do not accept the proposition that India is a multi-national society. The Indian people constitute one nation. India has expressed through her civilization over the ages, her strong underlying unity in the midst of diversity of language, religion, etc. The affirmation of India's nationhood after a long and historic confrontation with imperialism does not brook any challenge.

Nehru's 'unity in diversity' as the essence of Indian history (both civilization and nation – no distinction being made between the two) is the ideological basis of this position held by the Indian state. This 'unity of diversity' is not the same as the *Hindutva* (Hinduness) concept as a Hindu state, but is only a liberal 'assimilationist' view of the same Indian nation as the product of an Indian culture that was identified as a Hindu culture. Instead of an openly aggressive identification of the Hindu as Indian, Nehru's is a more, shall we say, Brahminical concept: it allows for the existence of various traditions, but insists on their subordination to an over-arching, Brahminical framework. It is only within this framework that even the 'secular' Nehru can identify Shivaji as the symbol of '*a resurgent Hindu nationalism*' (emphasis added). Since Nehru in *Discovery of India* (1946) certainly was not considering

Maharashtra as a nation, Shivaji could only have been upheld as a symbol of 'Hindu Indian nationalism'.

Benedict Anderson (1983, p. 19) makes the important point that nationalism has to be related to 'the large cultural systems that preceded it, out of which – as well as against which – it came into being'. Out of the fragmentation of the old sacred community arose the nations of Europe. According to Anderson, it was not the bourgeoisie's market drive that by itself created nations: 'In themselves market-zones, "natural-geographic" or politico-administrative, do not create attachments' (1983, p. 55). What was important were the vernaculars. These vernaculars replaced the old sacred languages – first, during the Reformation in providing direct access to the 'spiritual world' and, secondly, as state languages.

The spread of common vernacular languages in Europe promoted homogenization, a homogenization required by capitalist production. The quintessential bourgeois state of post-Revolution France aggressively promoted Parisian French as *la langue français*, in opposition to the other regional variants, which were termed as *patois*, regional dialects with no literary status. (Compare the almost parallel promotion of Khari Boli as 'Hindi', as opposed to such 'dialects' as Avadhi or Bhojpuri.)

These new state languages were very different from the old administrative languages. The 'old administrative languages were just *that*: languages used by and for officialdom for their own inner convenience. There was no idea of systematically imposing the language on the dynasts' various subjects' (Anderson, 1983, p. 45). But the new official languages were to be the instruments of integration of the communities, through 'print-capitalism' as Anderson calls it.

The choice of a particular vernacular as the official language had many consequences, chief of which was that it gave undoubted advantages to those people whose vernacular was adopted, particularly in empires where more than one language was spoken. For instance, in the Hapsburg Court of the Austro-Hungarian Empire, German was chosen as the official language. This provided an immediate advantage to German speakers as against speakers of all other languages in the Austro-Hungarian Empire.

Anderson points out that one would expect 'a self-conscious nationalism to arise last in each dynastic realm among the native-readers of the official vernacular' (1983, p. 76). This is correct, but only in a way. In India, for instance, 'self-conscious nationalism' seems absent in the Hindi belt, as compared to the other language areas, where self-conscious nationalism (or regionalism, as it is referred to in Indian political analysis) is well established. In the place of such a self-conscious nationalism, what exists is a self-conscious pan-Indian nationalism, parading as the Indian nation. Hindi nationalism defined as Indian nationalism provides definite advantages to Hindi speakers, and one can see why a separate Hindi nationalism is unlikely to arise, or may arise last of all in the Hindi realm. But is it correct to say that there is no Hindi nationalism, or rather, that Hindi nationalism passes as 'Indian nationalism'? Likewise, when we come to the role of religious communities, we will see that

'Hindu nationalism' identifies itself as Indian nationalism in opposition to the supposed *anti-national communalism* of the other religious minorities.

In Europe in the transition from civilization to nations, language played the crucial role – of course, brought alive by the drive for markets. The drive came from various capitals, as these came to existence in various regions. The bases to which these capitals related were the pre-capitalist cultures, out of which have been fashioned national cultures, identified largely with their specific languages.

The next phase of nationalism Anderson calls 'official nationalism', as against the earlier 'language nationalism'. This was characteristic of Czarist Russia (and later the Soviet Union), where there was a conglomerate of people speaking a number of different languages. The official nationalist policy consisted of Russifying the non-Russian areas of this empire. Language, with all that it connotes for culture, was the instrument of Russification, which was forced on the non-Russian European republics like the Baltic republics. And also on the Islamic republics, where 'The Soviet authorities, first "enforced" . . . an anti-Islamic, anti-Persian compulsory romanization, then in Stalin's 1930s, . . . a Russifying compulsory Cyrillicization' (Anderson, 1983, p. 48).

Here too, the drive for 'official nationalism' came from the rapid development of a Russian, large-scale bourgeoisie in Russia. It was able to use the other regions of the Czarist empire as colonial markets, and to prevent the growth of competing bourgeoisies in these colonies (which does not seem to have taken place in the ramshackle Austro-Hungarian Empire) and the later development of a new form of large-scale bureaucrat capital of the 'Soviet' variety.

2.2   Homogenization and Centralization

Important in this transition from civilization to nations is the process of homogenization that produces the national cultures, differentiated in Europe mainly on the basis of language. Homogenization is the special drive of the capitalist process, wherein even the promise (indeed, often not much more than promise) of equality is premised on a homogeneity, a potential homogeneity of everything other than access to capital. While feudal systems, in one way or another, regulate the access which various classes of subjects have to types of knowledge and consumption of various goods, capitalism places no such prior restrictions. The only restriction it recognizes is that of access to capital – in theory. In practice, the homogenization required by capital disadvantages not only those possessing lesser or no amounts of capital, but also, within the same nation-state, those belonging to cultures other than that of the dominant tradition, which is defined as the 'national culture'. The only way in which those (non-capitalists) of other cultural traditions can hope to compete on equal terms with members of the dominant tradition, is by recognizing the 'inferiority' of their own cultural system and accepting 'assimilation' into the 'national' cultural system.

As Samir Amin (1980, p. 31) expresses the necessity of homogenization: cultural life is not some mysterious, unfathomable domain, but is rather the way in which the utilization of use values is organized. Therefore, the homogenization of these use values through their domination by generalized exchange value must tend to homogenize culture itself. However, this tendency is held in check by the effects of unequal accumulation.

Along with homogenization of national culture, the capitalist state also brought centralization of economic and political powers. The state has always had an important role to play in capitalist accumulation. Among other things, it has been the instrument for centralizing the use of the surplus produced in a variety of spheres and modes and turning it into capitalist account. Competing capitals, or aspiring capitalists, necessarily have to try to control the state (their own nation-state) in order to further their own capitalist accumulation.

Consequently, just as centralization of surplus and homogenization of national culture unite the capitalists and the non-capitalist classes of the dominant (ruling) nation, so can resistance to centralized extraction of the surplus and to homogenization unite the capitalists and the non-capitalist classes of the disadvantaged (ruled) nations.

2.3 Indianness

Turning now to India, the fabled unity of India ('unity in diversity') was a civilizational unity. As Ravinder Kumar puts it, 'the political identity of the Republic of India should be defined in terms of a "Civilization-state" rather than a "Nation-state"' (1989, p. 39). Or, to cite Rajni Kothari (1989, p. 81): 'We need to remember that the essential identity of India is cultural, not political or economic. It is one civilization that has withstood various vicissitudes and still endured, largely because of its basic identity being cultural. It never had a political center except very recently'.

From a civilization an attempt is certainly being made to fashion one single nation. Crucial factors which have led to such an attempt include the long period of military-administrative unification of British colonialism and the growth of a large bourgeoisie serving an all-India market, which led the anti-colonial movement.

The drive for 'Indianization' comes from big capital. However, the base to which it relates is that of the pre-capitalist culture out of which an 'Indian' culture is sought to be forged. The basis of this 'Indian' culture has been defined as the upper-caste Hindu culture of the Hindi belt. It may be labeled *Ram Rajya*, but this is not a revivalism that seeks to bring back some values of the past. Rather, this is a use (even a creation) of the past to define something new – a uniform, homogeneous national culture centered on the cultural tradition of *Bharatvarsha* and *Aryavarta*.

The attempt to fashion a homogeneous national culture out of a civilization has led to a clash of cultures and of languages. Other cultural traditions, whether of different religious communities or language groups, are by definition not part of this 'mainstream'.

In a feudal situation, various religious communities could coexist and there was not necessarily an attempt to force a uniform cultural existence on all. Not all the religious communities had equal rights, however, and within the religious communities as well caste stratification defined unequal rights very rigidly. The 'togetherness' that Rajni Kothari finds in Indian civilization (1988) was based on *discrimination* between religious and cultural groups and between castes, and not on any concept of democracy or equality.

But, unlike in the Semitic religious communities, Indian civilization could accommodate different religious notions. Here again, accommodation was premised not on equality but on discrimination and the maintenance of suitable boundaries and distances, with each tradition being allotted a place in the social hierarchy.

This accommodation based on discrimination, however, changes with the attempt to transform a civilization into a nation. The homogeneity of the all-India bourgeoisie (not to forget the imperialist bourgeoisie, who would in fact be quite happy with an even larger area of homogeneity) requires that accommodation be replaced by an aggressive monism.

This change occurs both in the Hindu religious community and in its relation to others. In fact, as Ravinder Kumar (1985) points out, prior to the 19th century, 'Hinduism was innocent of any notion like the *umma* in Islam; or the "community" in Christianity; which could form the basis of a cohesive social formation'. The Hindu community was to be geared to not just 'purposeful spiritual activity' but also to 'secular endeavor'.

The Hindu community was now called into action, not as one of the various Indian communities, but as *the* Indian community. It was not only the religious revivalists, but the modernizing reformists as well who identified the Hindu community with Indianism and patriotism. Even before Bankimchandra's *Anandmatt* (an important novel of the second half of the 19th century portraying Muslim rule as the cause of India's downfall), Ram Mohan Roy and Vidyasagar had defined nationalism as the prerogative of the Hindu community.

Not only had the 'modernistic' trends like the *Brahmo* or *Prathana Samajas* or the more secular movements of Young Bengal or *Vidyasagar* been entirely Hindu in composition; with few exceptions, they too had operated with a conception of 'Muslim tyranny' or a 'medieval' dark age (an assumption we meet with in Rammohan and among Derozians almost as much as in Bankimchandra) from which British rule with its accompanying 'renaissance' or 'awakening' had been a deliverance. (Sarkar, 1983, pp. 75–76)

Most important, the cultural artefacts of Hinduism – whether in dress or language or any other aspect of social life – were defined as the expressions of 'Indianness'. Steps were taken deliberately to create a Sanskrit-based, Hindu language, Hindi, as against the earlier composite language, Hindustani. The all-India bourgeoisie supported the cause of Hindi not because it developed from this language community (in fact it did not) but because this language provided the most suitable instrument for the homogenization it required.

There are two variants of this expression of Indianness. One is the overt religious concept of *Hindutva*; and the other is the 'secular' expression of Indianness as based on ancient Indian (identified as Hindu) culture. The first is a religious, the second a cultural concept; but both together relate Indianness to the tradition of what is now identified as Hindu civilization. It is because of this unity between the overtly religious and 'secular' expressions of 'Indianness' that an alliance could exist between Gandhi's *Ram Rajya* and Nehru's 'secularism'; likewise, that Rajiv Gandhi, the continuer of this 'secular' tradition, could declare that *Ram Rajya* was his goal too.

The key problem is to understand 'Indianness' in terms of 'the large cultural systems that preceded it, out of which – as well as against which – it came into being' (Anderson, 1983, p. 19).

In Europe 'the large cultural systems' were distinguished from each other on the basis of language and religion, Protestant and Catholic variations of Christianity. The nations that came into being broke up the feudal religious community on these lines. The separation of religion from the state, turning religion into a private matter, was to a greater or lesser extent achieved in the course of this transition. But the cultural traditions on which the new characteristics of 'nationness' were defined were still those of a single religious system.

For all the successes of European nation-building, we should note that the one competing 'large cultural system', that of the Jewish religious community, could never be integrated into the definitions of the various, homogeneous national characters. The existence of a separate Jewish community did not pose much of a problem in the feudal period. But a homogenizing capitalism could not tolerate such a large measure of difference, so the earlier policy of 'tolerance based on discrimination' (Avineri, 1981, p. 8) gave way to systematic, state-sponsored pogroms to force the Jewish community into accepting the Christian-based cultural system as a precondition for the bourgeois promise of equality. In the end, Jewish resistance was to lead to the 'final solution' of the Holocaust and the Jewish community fleeing Europe. Through the instrument of Zionism, the Jewish religious community turned itself into an Israeli nation supported by various imperialist powers, itself expansionist and as ruthlessly discriminatory against the Palestinians and other Arabs as the Christian nations of Europe had been towards the Jews.

In India the problems of the differences in cultural systems related to different language-nations (regionalism, in the now-standard Indian political vocabulary) are partly understood. But the intractable question is that of the relation between different religious cultural systems and corresponding religious communities, leading to the political problem of 'communalism'.

As pointed out earlier, whether on an overtly religious basis as *Hindutva* or *Ram Rajya*, or on a 'secular' basis of 'equal treatment to all religions' (meaning favored treatment to the Hindu religious community), Indianness has been defined with the upper-caste Hindu/Hindi cultural tradition as the core. The cultural traditions of religious communities, to the extent they insist on being recognized as different from this 'core', are deemed 'foreign' (particularly the Islamic community and its traditions) or, as in the case of the Jharkhandis and other tribal-related cultures, as merely quaint curios, worthy only of being exhibited in festivals and parades.

In analyzing upper-caste Hindu/Hindi nationalism, we must distinguish between two distinct phases. During the period of British colonialism this narrow Hindu nationalism had a double role. While it had the negative effect of increasing the divisions in Indian society, it also played a positive role in mobilizing the Hindu community (more correctly: the upper-caste Hindu community) in the struggle against British colonialism and its injustices. In the last phase of this struggle, as the British were being forced to prepare to hand over power, the divisive role of this Hindu nationalism became more prominent. Subsequently, with the 'transfer of power' this upper-caste Hindu/Hindi nationalism was to emerge as the oppressor nationalism, and ceased to have any positive character. To support this nationalism is now to support the oppression of the minorities, linguistic or religious.

Partly in reaction to a Hindu-based Indianness, and more importantly, reflecting the limitations of their own elites' bourgeois ambitions, the Muslim and Sikh religious communities have raised and supported moves for their own brands of capitalist homogenization. Hindu-based Indianness, as well as Islamic and Sikh fundamentalism, are all attempts at forging bourgeois nations and homogeneous national cultures out of the different existing religious cultural systems.

In the common currency of Indian politics, the building of a Hindu-based Indianness is 'nationalism', becoming communalism only when predicated not merely on a religious culture, but on religion itself. Sikh and Islamic fundamentalisms, however, are by definition only and always 'communal'. The first holds state power and uses it to repress resistance by other communities to the homogenizing pressures. Sikh bourgeois nationalism is fighting to establish its own state, where it can elaborate a Sikh version of a uniform national culture and existence. Islamic fundamentalism is no longer in a position to assert a nationalism in the form of a demand for a state of its own (as it did in the colonial period), other than in Kashmir – but there too it is being challenged by Kashmiri nationalism.

2.4 Oppressor and Oppressed Nationalisms

To say that Hindu-based Indianness, and the Sikh and Islamic reactions to this, are all bourgeois nationalisms is not to put them at the same political level. A distinction must be drawn between the oppressor nationalism of Hindu-based, Hindi Indianness and the other oppressed nationalisms.

The CPI(M) holds that it cannot decide which is the 'oppressor nation and which are the oppressed nations' (1972, p. 10). At the same time, it admits that there is 'inequality' between Hindi-speaking and non-Hindi-speaking nationalities. (p. 14). What does inequality mean, if it does not confer privileges on one group, on the nation of the language? Even the States Reorganization Committee was aware that the linguistic minorities were disadvantaged. Today, after the English-speaking salariat (to use Hamza Alavi's term), it is only the Hindi speakers who have advantages outside their own state, in the central government departments. So strong is the Hindi nation's self-identification of itself as the center of India, as the Prussia of India, that it is almost universally believed by the 'people of the language' that Hindi is the national language of India and the sole guarantor of India's existence as a nation. Privilege for one nation is the counterpart of disadvantage for the other nations; and privilege and disadvantage are expressed at the political level as relations of oppressor and oppressed.

Are the workers and other toilers of the Hindi nation(s) part of the oppressor nation? To the extent that they insist on retaining for the Hindi nation(s) the privileges that flow from Hindi's status as the only vernacular official language at the pan-Indian level, to that extent they identify themselves with the oppressor nation. Membership of a toiling class does not automatically entitle a person to be regarded as not being part of the oppressor nation. Of course, the toilers, or their children who aspire to become part of the salariat, get little more than emotional (cultural) satisfaction out of being part of the oppressor nation. It is the Indian big bourgeoisie that alone really benefits from maintaining the other nations in relations of bondage. But, for the toiling classes not to be counted as part of the oppressor nations, they would have to take a political stand against the continuation of national privileges and national oppression.

One of the crucial internal weaknesses of Hindu nationalism is that it is based on a religious/cultural system that predicates the existence of and justifies a caste system. It is no accident that Gandhi's *Ram Rajya* included the continuation of the caste system, while accepting the need for modifying the worst excesses of untouchability. This close connection between Hindu culture and the caste system makes the cultural symbols of Hinduism that much less attractive for the lower castes, particularly the *dalits*. But this problem of contradictions within a socio-cultural system must be distinguished from the problem that arises from treating cultural systems as 'others', as something 'foreign'. The first is a contradiction that gives rise to the anti-feudal struggle, the second is a contradiction that gives rise to the national struggle.

Perhaps we can begin to understand the forces of 'communalism' and 'regionalism' in India when we relate the bourgeois nation-building process to the different 'large cultural systems' that have evolved in Indian civilization. These large cultural systems are differentiated on the bases of both language and religion. While the pan-Indian monopoly bourgeoisie which inherited the colonial state has chosen the upper-caste Hindu/Hindi cultural system as its base for 'Indianness' (not because all of them necessarily belong to this cultural system, but because this is the most effective instrument for a homogenization of the pan-Indian market), the smaller bourgeoisies, or aspiring bourgeoisies, have chosen their own large cultural systems, differentiated from the first on the basis of religion and/or language, with which to oppose the hegemony of the all-India monopoly bourgeoisie.

Consequently, widespread communalism in India cannot be understood without reference to the national question in India. The attempts to fashion an India on the basis of an upper-caste Hindu/Hindi cultural ethos, and the resistance of the linguistic and religious minorities to this attempt, are the chief factors behind communalism. Just as the religious wars of Europe preceded and were part of the process of formation of European nation-states, so is communalism in India related to the various attempts at nation-building. This comparison was made explicit in Dumont's analysis of communalism in India: 'It seems, from the present reflection, that those "religious" wars in Europe signalled the same transition that has produced communalism in India, and that there is thus some parallelism between the two phenomena' (1988, p. 318).

2.5   Categories of Analysis

The cultural homogenization of constructing a nation, and the economic drives of a uniform market and centralization, have their counterpart in the seemingly secular approach that refuses to accept the category of communities as analytically valid. The proponents of Nehruvian secularism and of the 'Left' alike ignore the existence of communities. For both there are only the imperatives of development and the necessary centralization. What development means in this sense is the development of the pan-Indian bourgeoisie. The key words then are 'efficiency', centralization, and 'development'. This ideology is very much in the interests of the big bourgeoisie, who do not want to make any concessions of economic (and political) territory to other bourgeoisies or aspiring bourgeoisies. Capital has no measure other than that of profit. What we then get is an alliance of steel and the sacred cow, with the 'Left' ensuring that the working class is also party to this alliance.

An analysis blind to the real existence of communities is not thereby neutral concerning these communities. A community-blind category is not community-neutral. It would be community-neutral if there were only one community – in which case there would be no problem at all. It could also be community-neutral in the highly improbable situation that the various communities were equally matched in their access to the state, capital, and so on. But in the actual situation of an existing dominant community, and of a pan-Indian bourgeoisie that comes overwhelmingly from this community, a community-blind approach would only help strengthen the monopoly of the Hindu big bourgeoisie. In reality, those advocating this approach end up in the same position as the proponents of *Hindutva*, who also want the oppressed minorities to forget the existence of communities. For the Hindu big bourgeoisie, the Hindu 'nationalists', the Nehruvian secularists and the so-called Left as well – for all of them there is only one Indian community, one and indivisible. It is instructive to note that just as the plainly revivalist Dr Sampuranand refused to allow any place for Urdu as an official language in Uttar Pradesh, in the recent controversy the 'Marxist' Namvar Singh also opposed according Urdu any official status, on the ground that this would be tantamount to recognizing 'communalism'.

Categories of analysis blind to the existence of certain social groups are not neutral among these groups but actually strengthen the dominant group in the particular sphere being analyzed. Neo-classical economics, bourgeois economics par excellence, is notorious for analyzing the market as a neutral venue of interaction between buyers and sellers. It ignores, for instance, the real difference between the buyers and sellers of labor power. An analysis blind to the existence of classes is certainly not neutral between classes. Similarly, an analysis that ignores the caste divisions of Indian society, in the name of 'overcoming casteism', is certainly not neutral between castes. Such an approach would only strengthen the domination of the upper castes. Finally, if policies are blind to the existence of gender, would they not therefore strengthen male domination? (The above analysis owes a lot to Nancy Hartsock, 1987, who argues that gender-blind categories are not gender-neutral.)

2.6   Role of Non-Capitalist Classes

Though we talk of bourgeois nationalism, and it is usually some form of bourgeoisie that leads these national struggles, this does not mean that other classes have no particular interests of their own in the struggles of oppressed nations. The twin problems of centralization and homogenization have to be confronted in the attempt to build a more democratic order.

The intelligentsia obviously plays an important role in the articulation of national feelings. The development of the vernacular languages, and the connection with these being official languages, is an essential part of any national movement.

Peasant movements too have been important in developing nations. Historically the Sikh movement of the Punjab peasants and that of the Maratha peasants against the Mughals have both been important in the development of these nations. Even today, peasant questions can be reflected in and through national movements.

In the case of the Sikhs, for instance, the problems of the peasantry are articulated through the form of Sikh nationalism. In Punjab, the peasants are basically Sikhs. Traders and merchants are largely Hindu; while the state, which is a necessary party to the agrarian situation, is that of the Hindu big bourgeoisie. Declining margins of the peasantry and the problems of unequal exchange through high prices of industrial commodities and relatively low prices of agricultural commodities easily become identified as a problem of exploitation of the Sikh peasantry by the Hindu, pan-Indian big bourgeoisie. Of course, the imperialist hand behind the Green Revolution and the present crisis of Punjab's agriculture is not so visible, operating as it does mainly through the mediation of New Delhi. The Sikh movement has, thus, remained at the level of appearance and not gone beyond it to confront the reality of imperialist control and exploitation. But the task of making this transparent will itself be served by removing the veil of New Delhi, and letting Punjab deal directly with imperialism and confront it.

The centralization of surpluses – which means making them available to the pan-Indian big bourgeoisie (who are linked with various imperialist trans-national corporations) for accumulation – operates through many structures. One, mentioned above, is that of unequal exchange between agriculture and industry. Another is that of the banking system, which gathers up deposits from agriculturalists and other classes, and places them at the disposal of the big bourgeoisie for investment. In Punjab, for instance, the proportion of advances to deposits of the banking system is some 43% (Jacob, 1990, p. 120) – the remaining 57% is then available for use by the big bourgeoisie else-where. Of course, of the portion advanced in Punjab itself, a large part goes to the pan-Indian bourgeoisie operating there. Other financial institutions also gather up various deposits, investment in shares, and the like. The various gov-ernment-owned financial institutions represent the major source of funds for capital accumulation by the pan-Indian big bourgeoisie. Consequently, bureaucratic connections are necessary to become and remain part of the big bourgeoisie. Another necessary factor is a foreign connection, in the form of technical or some other collaboration with a transnational corporation.

In Kashmir, too, the appropriation by the Indian state of the surpluses pro-duced by Kashmiri peasants and artisans is easily identified as the national exploitation of Kashmir by the Indian bourgeoisie. In the case of petty bour-geois, employees, and professionals as well, there is an identification of the discrimination they face as indicators of national exploitation.

The problem of cultural deprivation is also felt more keenly by the middle and lower classes – as compared to the upper and upper-middle class, the bourgeoisie, which is usually much more culturally mobile and 'cosmopoli-tan', easily adapting to the dominant cultural system. This is seen in the case of language as well. Children of peasants, who would almost invariably study in their own vernacular, have less access to jobs and professions than the children of urban traders and businessmen, who tend to have greater famil-iarity with the 'national' languages. Consequently, while the Tamil upper classes may champion the cause of English as against Hindi, this does not much help the mass of Tamil people. Those who study in Tamil cannot be as mobile as those who study in English. Further, the development of Tamil, as of the other vernaculars, is held up by the fact that higher education and official work is conducted either in English or in Hindi.

Therefore, it is not the bourgeoisie alone that is interested in ending national exploitation and subjugation. In fact, as both Punjab and Kashmir exhibit, the peasantry may be more determined and militant in its struggle against national exploitation. The top section of Sikhs, comprising the big capitalist landlords and middle bourgeoisie, have tended to support the more compromising policies of first the Barnala and then the Badal group of the

Akali Dal (the relatively moderate party of the Sikhs). By contrast, Bhindranwale (representing the militant, extreme stream of Sikh politics) had a solid base among the middle peasantry.

2.7 Minorities

The manner in which differences in cultural systems are stressed, by both Sikh and Islamic fundamentalism, shows that, while opposing homogenization into the dominant cultural system of Hindu 'Indianness', their proponents in turn wish to set up other homogenizations based on Sikh or Islamic religio-cultural systems. In this respect they are no different from the proponents of Hindu 'Indianness'. And, because the objective is, during and after the liberation struggle, to set up new homogenizations, we may say that these are forms of bourgeois nationalism. It is the new or aspiring bourgeoisies who again require uniform national cultures. Inevitably, such bourgeois nations, now oppressed, will oppress their own minorities as they gain state power. Various questions of exploitation and oppression internal to the nation are not solved simply by ending the state of that nation's oppression. Not only that: the national question too, in general, does not end with overcoming one level of national oppression; that of the minorities still remains.

While the problem of the minorities is usually seen as a cultural problem, and while there is a cultural dimension to this problem, it really concerns political and economic power. The existing system of elections does not allow for population-wise proportionate representation of, say, the Muslims, who are not geographically concentrated. Discrimination in entry into the bureaucracy and into the bourgeoisie is commonplace. Even someone not given to exaggeration like K.F. Rustamji (1990) refers to 'the inability of the minorities to get loans, licences and project approvals'.

In fact, liberal assimilation policies of the secularists would allow for a measure of cultural autonomy, while of course retaining the identification of upper-caste Hindu/Hindi culture as the central core of the nation's culture. Further, as seen in the episode of the Muslim Women's Bill, special forms of oppression might be allowed within the minority community. Anything may be allowed, short of an actual sharing of political and economic power – perpetuating the situation analyzed by Gellner (1983, p. 1) where 'ethnic boundaries within a given state . . . separate the power-holders from the rest' – i.e. the minorities.

Forms of decentralization of economic and political power to the national 'minorities' have to be considered. But such decentralization will be of a very limited kind under the rule of capital. The ruling class is the capitalist class, or more correctly, a combination of imperialists, big bourgeoisie, and landlords. When even politically sovereign states can be subverted by imperialist capital, there remains very little scope for a decentralization of power within the boundaries of a capitalist state. For example, in sovereign Bhutan, Indian capital is able to operate as lessees of nominally Bhutanese enterprises. Likewise, restrictions on ownership of property by outsiders in Kashmir or Manipur do not prevent Indian capital from again leasing enterprises in these regions.

To point out the limited effect of such restrictions is not to argue for their being abolished. In fact, it is Indian capital that would prefer a situation with no restrictions at all to its operations throughout South Asia. Not all imperialist powers are necessarily in favor of free trade and free export of capital. At any time, rising imperialist capital would prefer a situation with no restrictions in any part of the world; conversely, declining imperialist capitals would be very strongly protectionist in trade and would try to use political influence to maintain exclusive areas for capital export.

If, within the states that might be formed by presently oppressed nations, some form of decentralization would have to be attempted, then why should such decentralization not be attempted now, within the confines of the pan-Indian state? Such measures could certainly be demanded and attempted. But the manner in which the demand of the oppressed nations develops will depend very much on the history of oppression and suppression. Had there been a federal structure of the type envisaged by the Anandpur Sahib Resolution (with the federal government handling only defense, foreign affairs, currency, and communication), had there not been an Operation Bluestar, nor the 1984 massacres of Sikhs, had there not been the massive rigging of elections in Kashmir – it is conceivable that in such a situation, a new type of decentralization might have been tried out. But in all such matters, it is what the oppressed nations want that must form the starting point of discussion. The right of separation or secession has to be part of any scheme for political equality. If, as a result of a history of suppression and deprivation, an oppressed nation demands separation, there can be no question of this not being accepted.

2.8   History of Oppression and Separation

The growing separation between the Hindu and Sikh communities in Punjab has a long history. When the Sikhs agitated to regain control of their *gurdwaras* (temples) from Hindu *mahants* (religious leaders), Hindus at large in Punjab supported the corrupt and cruel *mahants* against the Sikhs. In Punjab (as in north India as a whole) language was inextricably mixed with religion. While Muslims wanted to retain Urdu as the official language, the Hindus were for Hindi. The Sikhs alone stood for the vernacular Punjabi as the official language. The Arya Samaj, with its doctrine of *Aryavarta* (the abode of the Aryas), *Aryadharma* (Hindu Vedas) and *Arya Bhasha* (Hindi language), was the main vehicle of the Hindu offensive in Punjab, and it was closely linked with and supported by the Hindu trading communities and Hindu middle classes (Banga, 1989, p. 217).

After 1947, the Arya Samaj and Congress both campaigned among Hindus to have not Punjabi, but Hindi, declared as their mother tongue. Sikhs were again left alone to defend the vernacular and canvass for it to be made the official language. When the question of linguistic nations led to the reorganization of states, Punjab was very deliberately not reorganized as a Punjabi state, because that would have left the Hindus in a minority there. In fact, the States Reorganizing Committee, set up to constitute the provinces of India on a linguistic basis, refused to recognize Punjabi as a distinct language (Jacob, 1990, p. 90). This echoed the manner in which, in the 19th century, it was denied that Sikhism was a distinct religion and the Sikhs a distinct community. Again, it was the Sikhs who not only spearheaded but virtually comprised the movement for recognition of the Punjabi language and formation of a Punjabi state. The spread of Hindi also meant Hinduization, as the texts invariably drew examples from Hindu myths and the Hindu cultural tradition. All-India Radio Jalandhar's station replaced, no doubt deliberately, words of Persian origin (which form up to 30% of the vocabulary in Punjabi) with Sanskrit-derived words.

In contemporary times, the divisions between the Sikh and Hindu communities in Punjab can be seen in the opposite reactions to important political developments. Hindus rejoiced at the fall of the Golden Temple to the Indian Army in 1984 and at the assassination of Bhindranwale, while the very same events pushed the Sikhs further away from New Delhi and India. Conversely the assassination of Indira Gandhi was certainly seen by Sikhs as a welcome act of revenge, while Hindus, particularly in the Hindi belt, saw it as a betrayal of the 'nation', and retaliated with large-scale massacres of Sikhs. If Sikhs demand the withdrawal of central para-military forces, Hindus will want their retention. If Sikhs want elections to be held to the Punjab Assembly, Hindus and the all-India parties will be scared that elections will bring into being a Punjab Assembly that could well, as a body, pass a resolution in favor of Khalistan, an independent Sikh state.

Oppression and suppression have a history: they are not constant or given. That the Indian ruling classes cannot follow any other policy than military suppression, when their forced assimilation is resisted, is another matter. But it is through history that communities and nations come into being. They do not come fully developed into the world; nor are they by any means unchanging, historical givens.

As pointed out earlier, decentralization under the rule of capital is bound to be very limited. A real decentralization, one that would not impose homogeneity, would be possible only with an economic system where profit does not dominate production, where exchange value does not dominate use value, where it is the associated producers who, at various levels, dominate production and decide on production on the basis of conscious decisions. This alone is the economic base that could allow cultural non-homogeneity to exist.

2.9 Struggles Within Oppressed Communities

From the above analysis it also follows that various struggles against the tenets of cultures that justify systems of inequality and oppression (as, for instance, concerning *dalits* and the status of women) have to be resisted within the specific community concerned. A universal approach to these problems will not do. Such a universalization of the 'uniform civil code', to override the tenets of other cultures, is exactly what the Hindu fundamentalists demand today.

Instead, there will have to be a concretization of the questions within each particular community. Even while we uphold the right of nations to form their own states and uphold the rights of 'minorities', there also has to be a struggle within each community/nation against its exploitative practices and doctrines.

Two important examples of struggles within the oppressed nations/communities concern *dalits* and women. Caste is such a pervasive feature of the South Asian social structure that, even in the case of egalitarian doctrines such as Sikhism or Islam, communities nevertheless are caste-ridden.

In the Sikh community too *dalits* occupy the lowest rungs of the social hierarchy; they rarely own land and, in accordance with the prescriptions of Manu (the first codifier of Hindu law), they perform menial services. For example, collecting cowdung is the exclusive preserve of *dalit* women, who, on pain of 'social boycott' (*nakabandi* – not being allowed to use the landowners' fields or common lands to ease themselves) have to perform this job. *Dalits* have separate settlements, called *chammariya* in some parts, and even separate *gurdwaras*. A struggle for ending the caste system, linked with the redistribution of lands of the landlords (whether feudal or capitalist) to the agricultural laborers and poor peasants, would certainly have to be a part of any democratic struggle of the Sikhs.

Early in 1991, the Sikh movement and the militants had some success in insisting upon the use of the Punjabi language in all official work in the state. However, it is no accident that along with this, the militants (or some sections of them at least) have also begun to impose dress codes, particularly for Sikh women.

Similarly, the process of the Shah Bano case and the Muslim Women's Bill (1986) showed that, while the attempt to use the Indian state as an instrument for change within the Muslim community was self-defeating, at the same time, the struggle to oppose the particular forms of oppression and subjection of Muslim women had to be fought within the oppressed community – a struggle that would invariably also have to be a struggle against that particular tradition.

This is the kind of combination of struggle against national oppression with a democratic struggle within the oppressed nation that cost Rajani Thirangamma her life in Jaffna.

2.10   Fundamentalism

The self-identification of a community in terms of its personal laws – which here essentially means different ways of oppressing women – is a poor substitute for an effective wielding of power by the community, or its dominant class. Why then does a community seek such an illusory self-identification? Whether it is a matter of Islamic nations facing the onslaught of imperialism, or of the Muslim community in India facing the onslaught of the sub-imperialist, pan-Indian big bourgeoisie, the choice of identificatory marks in the community's dealings with the internal structures is a sign of its inability to deal with actually existing relations with the external, dominant powers. Fundamentalism, then, signals the inability of the community to deal with the relations imposed on it by external powers, namely by different forms of imperialism and sub-imperialism.

Take the example of Sikh fundamentalism in the Punjab. It is only part of the truth that New Delhi's sub-imperialism is responsible for the crisis of Punjab's agriculture and economy. Behind the crisis of the Green Revolution lie imperialist finance capital and the multinationals. Nowhere is this challenge posed by imperialism part of the self-identification of Sikh fundamentalism. While Punjab's agriculture could provide the base for the development of a self-reliant industrial structure, one that could challenge the domination of imperialism, Sikh fundamentalism is seeking a solution to its current problems by going even further along the path of dependency on imperialism, insisting on the 'Pepsi' kind of neo-colonial industrialization.

The important thing about the existence of illusions is 'the demand to give up a state of affairs which needs illusions' (Marx, 1844, p. 176, emphasis in original). In the case of the Muslim community in India, the powerlessness of any section of the community, the continuous attacks on the besieged community, constitute 'the vale of tears, the halo of which is religion' (1844, p. 176, emphasis in original). It is the real condition of existence in a vale of tears that will have to be changed if illusions are to disappear.

Fundamentalisms – whether Muslim or Sikh – are attempting to set up not individual spiritual authority, but theocratic authority. Political action by a community does not, however, require that there be a collective religious authority. On the other hand, full participation of individuals in a liberation movement requires that the individuals can take their own, conscious decisions about participating in or abstaining from a movement. The free development of the individual depends on the community, just as the free development of the community depends on the conscious actions of the individuals who comprise it.

2.11 Whose Compulsions?

In any solution to problems, various compulsions will be in operation. Conflicting compulsions reflect the conflicting interests of various classes. From whose compulsions should we then start out in our search for solutions? From the compulsions of the Indian big bourgeoisie, implementing a policy of building a centralized state and a homogeneous India on an upper-caste Hindu/Hindi cultural base? Or from the compulsions of those who are oppressed by this policy?

For whose progress or development is greater centralization of economic powers required? A uniform pan-Indian market, with a centralized state, capable of centralizing the utilization of surpluses – this is necessary for the Indian big bourgeoisie. Imperialism too would have no problem in dealing with one centralized state and one uniform market, rather than with a number of states with varying conditions and regulations and heterogeneous markets.

In the period before the transfer of power in August 1947, the Hindu big bourgeoisie opposed federal schemes that would give Muslim-majority provinces a measure of autonomy. The big bourgeoisie, in the model set out in the Bombay Plan, required a strong central government, one able to deploy surpluses as required by the class interests of the big bourgeoisie. A federal scheme would have helped the formation of various competing, national (regional) bourgeoisies. Big business preferred the separation of Pakistan to some such federal scheme:

> . . . the support extended by Hindu big business to Pakistan is less paradoxical than it would appear at first sight. For the separation of the Muslim-majority provinces facilitated the creation of a fairly centralized State, endowed with a strong central government. (Markovits, 1989)

When Pakistan came into being, the dominant bourgeois–bureaucrat class there imposed the same kind of centralization as in India. This attempted centralization and homogenization, economically exploiting the surpluses produced in East Pakistan while suppressing, among others, East Pakistan's distinct linguistic, cultural system, was resisted by the (Muslim) Bengali nation. In the end, it was to lead to the creation of Bangladesh.

Pakistan is as much multi-national as India is. Religion alone could not form the basis of a nation, and the cultures of a religious community could be broken up by language, as is evident from the breakaway of Bangladesh from Pakistan in 1971. And conflicts based on ethnicism and nationalism persist.

That a religious community can be broken up by language is apparent in the case of the Azeris. The Muslim Azeris, who live in both Azerbaijan and in Iran, have a Turkic language quite distinct from Persian. Despite their shared religion, fear of Azeri nationalism within Iran has restrained the Islamic fundamentalist regime in Teheran from all-out support to Azeri nationalism in the Soviet Union.

The 'Left' parties in the Indian Parliament, though they may talk of more powers for the states, oppose the kind of decentralization envisaged in the Anandpur Sahib Resolution of the Sikh political party Akali Dal. H.S. Surjeet of the CPI(M), for instance, writes in *The Hindu* (1989):

> Even on the issue of autonomy, once the issue was under discussion nobody who really desires national unity and progress would be able to argue that merely four powers (defense, foreign affairs, currency and communication) are sufficient for the center. This is not in the interests of the States which are constituents of the federal set-up.

Would smaller economic units (or even sovereign states) mean less development or progress for the units concerned? Development, even in the bourgeois sense, depends not on the size of the population, but on the rate of its internal transformation. The more complete the liquidation of pre-capitalist relations, the faster would be the rate of growth. Also crucial here is the rate of deepening of capital and production, along with the increase of wages.

External markets played an important role in the formation of capitalism in Europe. But what was decisive was the degree of internal development of the capitalist class concerned. Some 'small' nations, but large economies, like Britain and more recently Japan, have played a dominant role in world industry. There is no necessary and direct relation between the size of nations and their rate of development. In fact, the existence of internal colonies may well reduce compulsions for the liquidation of pre-capitalist relations, thus inhibiting bourgeois development or progress.

Consequently, it is incorrect to refer to the necessity of 'development', without any adjectival qualifications, as the reason for supporting a large, centralized state. It is the Indian big bourgeoisie that requires a large, centralized state. Decentralization or even smaller sovereign states would help the growth of middle-level, regional (national) bourgeoisies. Such regimes are not in the interests of the Indian big bourgeoisie, however. A large, centralized state is also in the interests of the mobile intelligentsia and professionals, who supply the officials and intellectuals of the state. The Tamil Brahmins, for instance, are an excellent illustration of a salariat that staunchly defends the pan-Indian identity and state.

In terms of increasing the extent of democracy, it is necessary to bring about a form of national state that can reduce, if not eliminate, national oppression. Socialism cannot be conceived of without the largest measure of democracy; and the continuance of national oppression is inconsistent with democracy.

Another oft-cited compulsion besides 'development' is that of 'geopolitics'. For instance, Ashok Mitra (1989) writes on Kashmir: 'The activists currently on the rampage (in the Kashmir Valley) must be assured in advance that, should they come to win the poll, New Delhi would indeed permit them to hold the reins of administration in Srinagar.' But there is a rider – these young activists must not go too far; and they 'must recognize the importance of continental geo-politics'.

The doctrine of 'geo-political compulsions' is merely a way for a bigger, expansionist state to impose its will on weaker neighbors. 'Geo-political compulsions' are supposed to determine that the Himalayan ranges are necessary for the security of India and thus dictate that Kashmir must be held, even against the will of its people. Likewise, it is 'geo-political compulsions' that determine that the Indian state has the right to decide – or at least influence – policy in Nepal or in Sri Lanka, so as to safeguard the 'legitimate' security interests of the Indian state. What is at stake are not genuine security interests, but the interests of being able to dispose of labor surpluses and exploit markets. It is the pan-Indian bourgeoisie whose interests are involved in holding on to the Kashmir market and using its surplus.

Ashok Mitra is in good company in insisting that the Kashmir militants respect the (imperial) importance of continental geo-politics. It was Lord Curzon who first enunciated the imperial strategic doctrine for the Indian empire:

> India is like a fortress, with its vast moat of the sea on two of her faces and with mountains for her walls on the remainder; but beyond these walls, which are sometimes of by no means insuperable height, and admit of being easily penetrated, extends a glacis of varying breadth and dimension.
>
> We do not want to occupy it, but we also cannot afford to see it occupied by our foes. We are quite content to let it remain in the hands of our allies and friends, but if rivals creep up to it and lodge themselves right under our walls, we are compelled to intervene because a danger would thereby grow which might one day menace our security. . . . He would be a short-sighted commander who merely manned his ramparts in India and did not look beyond. (Quoted in Maxwell, 1970, p. 21)

So, the compulsions of both 'development' or 'progress' and of 'security' are the compulsions of the pan-Indian *haute bourgeoisie*. They are not the compulsions of those who seek to build a more democratic order. An essential component of such a democratic order is that there should be an end to national oppression, that there should be national equality.

One more factor always crops up in discussions on the national question in India: the role of imperialism and the disintegration of the Indian state. Imperialist pressures there certainly are on the Indian state. To what extent they are resisted depends on the strength of the ruling classes, and even more on resistance by the people of various communities and classes. The suppression of some communities and nations will only make it easier for various imperialists to operate in Indian politics; while a more democratically set order, whatever its size, based on voluntary acceptance by the communities and nations, will be able to resist imperialism more effectively.

Further, in any situation of conflict among the imperialists, there will always be some who will support national movements, while those who dominate the central state will oppose them. For instance, if US imperialism supported oppositional movements, the late-unlamented Russian social imperialism then supported both centralization in New Delhi and Indian expansion.

That one imperialist or another may support a national movement does not in any way negate the national movement. Did the support of Indian expansionism and Russian social imperialism for the Bangladesh liberation movement invalidate it? Given competition among various imperialist powers, it is inevitable that some imperialist or other will support any given national movement.

In framing policies for a more democratic order, where then should we start? From the 'compulsions' of the Indian big bourgeoisie, whether on the right of separation or on the extent of decentralization? Or, from the need to eliminate all forms of oppression of nations, religious communities, and the like? The choice of starting-point is all-important in determining the solutions one is willing to support. Accepting various 'compulsions' – compulsions that are essentially those of the Indian big bourgeoisie – will finally end up as a subtle defense of the existing order.

At the same time, supporting the right of separation is only the beginning and not the end of a solution. By contrast, to those bourgeoisies or aspiring bourgeoisies whose interests lie in wresting some economic space from the pan-Indian bourgeoisie, separation is the end-solution in itself. Having their own state, or a respectable share of it, is the sole aim of their struggle to create better conditions for their own accumulation. On the other hand, the removal of one form of national oppression can help in creating some more democratic conditions for the elimination of capital itself. At present, however, it must be emphasized that the struggle for a more democratic society cannot be delinked from the struggle to ensure the right of separation, and the economic and political rights of minorities.

The ultimate unity of the oppressed toilers can be possible only if they can supersede the philosophy of nationalism: but '*you cannot supersede philosophy without making it a reality*' (Marx, 1844, p. 181, emphasis in original). Only through realizing and struggling against nationalism, can the necessity of a new organization of the community begin to be realized by those who will need to create this new community of people.

References

Amin, Samir, 1980. *Class and Nation, Historically and in the Present Crisis*. New York: Monthly Review Press.

Anderson, Benedict, 1983. *Imagined Communities, Reflections on the Origin and Spread of Nationalism*. London: Verso.

Avineri, Shlomo, 1981. *The Making of Modern Zionism*. London: Weidenfeld and Nicholson.

Banga, Indu, 1989. 'The Growth of Hindu Consciousness in Punjab', pp. 201–207 in P.C. Chaterjee, ed., *Self-Images, Identity and Nationality*. New Delhi: Allied Publishers.

CPI(M), 1972. *National Question in India*. New Delhi: Odyssey Press.

Dumont, Louis, 1988. *Homo Hierarchiness*. New Delhi: Oxford University Press.

Gellner, Ernest, 1983. *Nations and Nationalism*. Oxford: Oxford University Press.

Government of India, 1984. *White Paper on the Punjab Agitation*. New Delhi: Government of India.

Hartsock, Nancy, 1987. *Money, Sex and Power*. Boston, MA: Northeastern University Press.

Jacob, T.G., 1990. *Punjab Crisis*. New Delhi: Odyssey Publications.

Kothari, Rajni, 1988. 'Class and Communalism in India', *Economic and Political Weekly*, Bombay, 3 December.

Kothari, Rajni, 1989. 'Cultural Context of Communalism in India', *Economic and Political Weekly*, 4 January.

Kumar, Ravinder, 1985. 'Gandhi, Ambedkar and the Poona Pact 1932', *South Asia*, June–December.

Kumar, Ravinder, 1989. *India: a 'Nation State' or a 'Civilization-State'*. Occasional Papers, New Delhi: Nehru Memorial Museum and Library.

Markovits, Claude, 1989. 'Businessmen and Partition of India'. Paper at Business History Seminar. Ahmedabad: Indian Institute of Management.

Marx, Karl, 1844. 'Contribution to a Critique of Hegel's Philosophy of Law. Introduction'. Reprinted in 1975 *Collected Works*, vol. 3. Moscow: Progress Publishers.

Maxwell, Neville, 1970. *India's China War*. Bombay: Jaico.

Mitra, Ashok, 1989. *The Indian Post*. 27 December.

Nehru, Jawaharlal, 1946. *Discovery of India*, London: Meridian.

Rustamji, K.F., 1990. 'Kashmir: the Need to Look Inwards,' *The Times of India*, 12 February.

Sarkar, Summit, 1983. *Modern India*. New Delhi: Macmillan.

Surjeet, Harkishan Singh, 1989. *The Hindu*, 25 December.

3

The Dynamics of Power: Military, Bureaucracy and the People

AKMAL HUSSAIN

3.1 Introduction

The available literature on the nature of state power in Pakistan has essentially examined how the state apparatus came to predominate over the political system. (Alavi, 1983; Hussain, 1990a; Jalal, 1990) Within the state apparatus, the bureaucracy and the military have so far been lumped together as co-sharers of the piece of the power-cake that has accrued to the 'state apparatus' as opposed to the political elites in civil society. The dynamics *between* the bureaucracy and the army, and the changing internal balance of power *within* the state structure itself have hitherto not been analysed. It would be useful to examine these dynamics, since the bureaucracy and the military are quite different institutions. They not only relate in differing ways to civil society, but, it can be argued, have in fact moved in opposing directions in terms of the nature of internal changes.

This chapter looks into the changing balance of power between the bureaucracy and military *within* the state structure. First, we examine the nature of the crisis confronting any authority that purports to govern. Next, *intra*-institutional changes, as well as inter-institutional changes with respect to the bureaucracy and military respectively are analysed. Finally, the role of the people is examined, as a factor influencing the power structure when the institutions of civil society have been eroded.

3.2 Economic Growth, Social Polarization and State Power

At the dawn of Independence in 1947, Pakistan's ruling elite consisted of an alliance between landlords and the nascent industrial bourgeoisie, backed by the military and bureaucracy. The nature of this elite conditioned the nature of the economic growth process. However, the latter in turn influenced the form in which state power was exercised. Economic growth brought affluence to the few, at the expense of the many. The gradual erosion of social infrastructure, endemic poverty and growing inequality between the regions undermined civil society and accelerated the trend towards militarization.

3.2.1 Economic Growth and Social Polarization

While the average annual growth rate of GNP fluctuated during the regimes of Ayub Khan, Zulfiqar Ali Bhutto, Zia-ul-Haq and Benazir Bhutto, the overall trend of growing poverty and social and regional inequality continued.

During the Ayub period (1960–69), the basic objective of Pakistan's development strategy was to achieve a high growth rate of GNP within the framework of private enterprise supported by government subsidies, tax concessions and import controls. Investment targets were to be achieved on the basis of the doctrine of functional inequality. This meant deliberate transfer of income from the poorer sections of society, who were thought to have a low marginal rate of savings, to high-income groups, who were expected to have a high marginal rate of savings. It was thought that by thus concentrating incomes in the hands of the rich, total domestic savings and hence investment could be raised.

This strategy was put into practice during the 1960s. But while income was transferred into the hands of the rich, they failed to increase their savings significantly – thereby obliging the government to increase its reliance on foreign aid in order to meet its ambitious growth targets. The growth process in Pakistan during this period generated four fundamental contradictions:

- A dependent economic structure and growth inflow of foreign loans (from USD 373 million between 1950 and 1955 to USD 2,701 million in 1965–70).[1]
- An acute concentration of economic power (43 families owned 76.8% of all manufacturing assets by the end of the 1960s) (Hussain, 1988, 1990b).
- Polarization of classes in the rural sector and a rapid increase in landlessness.[2] While the incomes of the rural elite increased sharply following the Green Revolution, the real incomes of the rural poor declined in absolute terms. Per capita consumption of foodgrains among the poorest 65% of Pakistan's rural population fell from an index of 100 in 1963 to 91 in 1969.[3] Similarly, according to a field survey, 33% of small farmers operating less than 8 acres suffered a deterioration in their diet. During the 1960s, as many as 794,042 small farmers became landless labourers (Hamid, 1974).
- Growing economic disparity developed between the various regions (Hussain, 1985).

These consequences generated explosive political tensions which not only overthrew the Ayub government, bringing in Yahya Khan's martial law, but also fuelled the secessionist movement in East Pakistan which ultimately resulted in the formation of Bangladesh.

During the Bhutto period, economic growth slowed down markedly. Industrial growth fell from an average of 13% during the 1960s to only 3% during the period 1972–77. Similarly, agricultural growth declined from an average 6.65% in the 1960s to a mere 0.45% in the period 1970–76.[4] At the same time, the nationalization of banks and credit expansion for financing loans to capitalist farmers and industrialists led to heavy deficit financing and an associated increase in the money supply. (Bank-note circulation increased from Rs 23 billion in 1971–72 to Rs 57 billion in 1976–77.) The sharp increase in the money supply during this period of virtual stagnation was reflected in a steep rise in the inflation rate: the wholesale price index rose from 150 in 1971 to 289 by 1975 (Hussain, 1988).

Although nationalization of industries and credit expansion enabled the Pakistan People's Party (PPP), then in power, to acquire the support of some of the urban petit bourgeoisie by providing jobs, licences and loans, the funds available were apparently not enough to enrich the entire petit bourgeoisie. In fact, the section of the lower middle class that did not gain from the PPP suffered an absolute decline in their real incomes due to the high inflation rate. It was this frustrated section of the petit bourgeoisie and the large lumpen proletariat stricken by inflation, that responded to the call by the Pakistan National Alliance (PNA), an electoral alliance between nine opposition parties, for street agitation in March 1977. Although the apparent form of the street agitation was spontaneous, it had been orchestrated and given political focus at key junctures by the PNA, which charged the government with rigging the elections. This organizational and coordinating function was performed by trained cadres of the Jamaat-e-Islami (party of the religious right), allegedly with support from the USA. The agitation was, of course, fuelled by the allegations that the PPP had rigged elections in several constituencies. The overthrow of the Bhutto regime and the subsequent hanging of the first popularly elected Prime Minister of Pakistan dramatically demonstrated the limits of populism within a state structure dominated by the military and the bureaucracy.

3.2.2 The Fragmentation of Civil Society

Each regime that has come into power in Pakistan has sought to legitimize itself through an explicit ideology. The Ayub regime propounded the ideology of modernization and economic development. The Bhutto regime sought legitimacy in the ideology of redeeming the poor (food, clothing, shelter for all) through socialism. It is an index of Zia's fear of popular forces, that he initially sought justification for his government precisely in its temporary character. If anything this was the ideology of transience – that he was there for only 90 days; and for the sole purpose of holding fair elections. It was this fear that impelled the Zia regime to seek (albeit through a legal process) the physical elimination of the one individual who could mobilize popular forces. It was the same fear that subsequently induced Zia to rule on the basis of military terror while propounding a version of Islamic ideology. Draconian measures of military courts, arbitrary arrests and public lashings were introduced. Thus the gradual erosion of the institutions of civil society brought the power of the state into stark confrontation with the people. Earlier in 1971, this confrontation had been a major factor in the breakup of Pakistan and the creation of an independent Bangladesh. Now a protracted period of martial law under the Zia regime served to brutalize and undermine civil society in what remained of Pakistan.

As the Zia regime militarized the state structure, its isolation from the people was matched by its acute external dependence. In the absence of domestic political popularity it sought political, economic and military support from the United States. This pushed Pakistan into becoming a 'frontline state' in America's Afghan war, and became an important factor in further undermining civil society.

The years between 1977 and 1987 saw a steady inflow of Afghan refugees into Pakistan and the use of Pakistan as a conduit for arms for the Afghan war. Two trends emerged to fuel the crisis of civil society:

- A large proportion of weapons meant for the Afghan guerrillas filtered into the illegal arms market.
- A rapid growth of the heroin trade. Powerful Mafia-type syndicates emerged to operate the production, domestic transportation and export of heroin. Many Afghan refugees, who had taken over a significant share of inter-city overland cargo services, also became integrated into the drug syndicates.

The large illegal arms market and the burgeoning heroin trade injected both weapons and syndicate organizations into the social life of major urban centres in Pakistan. At the same time, the frequent bombings in the North West Frontier Province during the late 1980s, because of the Afghan war and the weakening of state authority in parts of rural Sindh, served to undermine public confidence in the basic function of the state: that of providing security of life for its citizens. Under these circumstances it is not surprising that more and more people should begin seeking alternative support mechanisms in their communities to obtain redress against injustice and to achieve security against a physical threat to their persons and families. The proximate identity or group membership through which the individual seeks such security can be an ethnic, sub-religious, sub-nationalist or *biradri* (kinship) group. Civil society has now begun to become polarized along vertical lines. Each group – whether ethnic, sub-religious, sub-nationalist or *biradri* – has an intense emotional charge, as well as a high degree of firepower derived from the contemporary arms market.

3.2.3   The Crisis of Development

In the context of development, governments in Pakistan are faced with a crisis that has four features:

- Economic growth has been associated with poverty, and in some areas growing poverty. Almost 40% of the people are unable to obtain 2,100 calories a day per person. There has been impressive GNP growth (5.5%

annual growth rate during the Ayub period, 6.5% during the Zia regime, and just over 5% during the brief tenure of the Benazir Bhutto government). Yet, after 43 years, a substantial proportion of the population remains deprived of even the minimum conditions of human existence (Hussain, 1988). As much as 64% of the population lack access to piped drinking water. The percentage without 'safe' drinking water is probably larger, since piped drinking water frequently carries bacteria. The housing situation is so bad that 81% of the housing units have on average 1.7 rooms which are inhabited by on average 7 persons. Finally, the literacy rate of 28% is amongst the lowest in the world, and the standards of those few who make it to college are spiralling down at a dizzying pace.

The overall consequence of these features is a growing pressure on a fragile democratic polity. A significant section of the population perceives that there is nothing for them in this growth process – which becomes a factor in the resurgence of sub-national groups. Consequently, a new conflict may be emerging between centralized state structures and a polarized polity, associated with a heightened level of violence in society.

- The second element in the crisis is the rapid urbanization rate. Given current trends, the urban population is expected to double over the next decade and, what is worse, it is likely to be concentrated in large cities. With the prohibitive cost of providing basic services in large cities and the financial squeeze on the government, a growing proportion of the urban population would be deprived of even minimum civic services. Thus, the percentage of urban population living in unserviced localities (called *katchi abadis*) is expected to increase from today's 25% to 65% by the end of this century.[5] The level of social stress and associated violence may become difficult for any future government to handle. Thus, policies for slowing down urbanization and for increased investments in basic services are imperatives for sustainable development.
- The third element of the existing development process is rising debt. With existing levels of indebtedness, and government expenditure on unproductive purposes, an attempt to accelerate GNP growth substantially could land Pakistan with an intolerable debt-servicing burden. Latin America can serve as an example of what can happen when high growth rates are attempted with high levels of debt. The total debt in just four Latin American countries (Argentina, Brazil, Mexico and Venezuela) was over USD 282 billion in the early 1990s, or two-thirds of the outstanding loans of banks to all developing countries. When debt-servicing burdens in Latin America rose, the creditors enacted a squeeze which slowed down GNP growth to a point where real per capita income actually declined in some cases.

In Pakistan today the situation is not as acute as in Latin America. Yet, debt servicing as a percentage of foreign exchange earnings is already 25%. An alarmed IMF has introduced a credit squeeze which is already slowing down the GNP growth rate in Pakistan.

- The fourth feature is the rapid erosion of the natural resource base: the depletion of forests, desertification resulting from soil erosion and salinity, the rising toxicity levels of rivers due to untreated disposal of industrial effluents, while rising levels of air pollution are not only making the present hazardous, they also limit the possibility of escaping from the poverty trap in the future (Qutub, 1991).

Failure to devise a strategy capable of coming to grips with this development crisis has been an important factor in social polarization and the resultant difficulty in strengthening democratic institutions, particularly a democratic culture. The deepening of this economic and social crisis presents a challenge of governance to the three centres of power that purport to govern in Pakistan: the civilian political elite (through parliament and its executive authority), the bureaucracy and the military. The relative power that each of these protagonists is able to wield may depend on the effectiveness with which it can provide solutions to this crisis. In the next section, we will see how the balance of power within the state structure has shifted from the bureaucracy towards the military.

3.3   The Changing Internal Balance in the Structure of State Power

The changing relationship between the military and bureaucracy, the two vital elements of the state apparatus in Pakistan, can be understood in the context of three analytically distinct but interactive processes. These are in turn conditioned by the dynamics of Pakistan's security environment and its foreign policy priorities, particularly its relationship with the United States.

- Changes in the internal sociology of the military and bureaucracy, associated with changes in the social origins of officers in these two institutions.
- Changes in the professional quality of officers and the internal cohesion of the institutions.
- The balance of power between the state apparatus on the one hand, and such institutions of civil society as parliament, political parties, media and various fora of public expression, on the other.

In this section we will examine how these three processes have influenced the dynamics within and between the bureaucracy and the military. Over the past three decades, the social origins of both the bureaucracy and the army have shifted, from the landed elite to a wider base in the urban middle strata and the burgeoning class of rural capitalist farmers.[6] The latter class did include scions of some of the earlier feudal landlords who had transformed themselves following the Green Revolution of the late 1960s, when new, high-yield varieties made owner cultivation with hired labour an economically attractive venture. However, these capitalist farmers also included many rich peasant families who were able to move up the social scale by reinvesting the increased profits that became available from farming (Hussain, 1988, Part IV). While the change in social origins of officers in both these institutions has tended in the same direction (a broadening of the social base), changes in the level of professional competence and indeed the internal institutional cohesion have moved in opposing directions with respect to the bureaucracy and military.

3.3.1   Institutional Decay of the Bureaucracy

During the past 40 years, Pakistan's bureaucracy has undergone a gradual process of institutional decay. Perhaps the single most important factor here has been a sharp decline in the intellectual calibre of the civil servant, caused primarily by the collapse of academic standards at colleges and universities, and by the institutional failure to provide high quality in-service training. To make matters worse, the best products of even the present poor education system do not normally sit for the civil service examination, but the structure of the civil service remains predicated on the now-unfounded assumption that the 'intellectual cream' of society applies for and enters the service. Having entered the civil service, these poorly educated young officers face a future in which there is an absence of rigorous formal education to equip them professionally for the tasks they are supposed to perform.

Three institutions purport to provide a semblance for 'training' to the civil servant: the Pakistan Academy for Administrative Training, which gives courses to each crop of fresh entrants to the civil service; the National Institute of Public Administration (NIPA), which gives courses to officers at the middle stage of their careers (deputy secretary level); and the Pakistan Administrative Staff College (PASC), which gives training to senior officers, federal joint secretaries and heads of departments. In all three institutions there is a virtual absence of a high-quality faculty, and reliance is placed on invited speakers who lecture and then leave. Courses are so superficial and participant evaluation so soft as to pose no great intellectual challenge.

The decline in the intellectual quality of individual officers has been accompanied over the past two decades by an erosion of institutional decision-making mechanisms in the civil service. Political factions at various points in the political power structure interfere arbitrarily in a wide range of decisions – from transfers, promotions and dismissals of officers or judicial decisions by district commissioners on land disputes, right up to the issues of arrest of drug barons or approval of major projects. The integrity of institutional decision-making is often undermined by vested interests outside the civil service. This has resulted in increasing insecurity, corruption and on occasion demoralization of civil service officers. Such attitudes may have been reinforced by the large-scale dismissals of senior officers, sometimes on flimsy charges by successive regimes. For example, Ayub Khan dismissed 1,300 civil service officers in 1959 by a single order; then in 1969, 303 were dismissed by General Yahya Khan; during the regime of Z.A. Bhutto, as many as 1,400 were dismissed through a single order; and again in 1973, 12 senior civil service officers were unceremoniously removed.

At a structural level the CSP (Civil Services of Pakistan) was the elite cadre within the civil bureaucracy and its members inherited the ICS (Indian Civil Service) tradition. The CSP cadre remained dominant in the bureaucracy and indeed over national decision-making, right up to the end of the Ayub period. During the subsequent brief regime of General Yahya Khan, the dominance of the CSP began to be broken by the military authorities. Subsequently, the regime of Z.A. Bhutto further eroded the internal cohesion and esprit de corps of the CSP by a policy of 'lateral entry' into the service. This meant that individuals politically loyal to Bhutto, whether from various government departments or outside the bureaucracy altogether, could be appointed to key civil service positions. During the regime of General (later President) Zia-ul-Haq, the position of the bureaucracy within the structure of state power was rehabilitated. Zia gave greater confidence to civil servants by putting an end to the device of 'screening' civil servants which, during the regimes of Yahya and Bhutto, was like a sword of Damocles hanging over in-service bureaucrats, who could be dismissed or transferred at short notice. Senior bureaucrats now had relatively long tenures.

In the regime of Prime Minister Benazir Bhutto, new stresses were placed on the structure of the bureaucracy as a result of the growing political conflict between a PPP government in the centre and the opposition IJI (Islamic Democratic Alliance) government in Punjab, the largest province. The historically unprecedented contention for power between the federal and Punjab Provincial Government often took the form of manipulating individuals or groups of civil servants. The use of bureaucrats as instruments of the political power struggle between the Centre and the province was manifested dramatically in two cases.

The first concerned the federal government's decision to transfer to Islamabad five senior officers working in the Punjab Provincial Administration (the Inspector General Police, Superintendent Police, Information Secretary, the Additional Chief Secretary and the Chief Secretary in the Punjab). According to the federal government, these officials were misusing their power for the pursuit of political interests of the provincial government. The Punjab government initially resisted and then acquiesced to the transfer orders for four of the five officers. In the case of the Chief Secretary of the Punjab government, Anwer Zahid, the federal government's instructions to transfer him were successfully resisted by the then Chief Minister for Punjab, Nawaz Sharif.

The second case concerned implementation of the federal government's People's Programme for Development (PPD). This envisaged providing basic services to the poor at the grass-roots level, such as schools, drinking water, brick-paved village streets and drains. The federal government, which had also provided the funding, attempted to run a set of development activities which normally fell within the purview of the provincial government as one of their projects. The provincial government decided to resist implementation of the People's Programme for Development, on grounds that it was an attack on their authority. This conflict created surrealistic scenes of villagers building roads and drains with bricks, while the local deputy commissioner sent bulldozers to demolish the construction and arrested the workmen on charges of disturbing public peace.

The typical civil servant in Pakistan today is faced with formidable problems of poverty, social polarization, breakdown of law and order and erosion of infrastructure. He is presumed to be tackling these problems in an environment where often-conflicting demands from a still nascent political system are impinging upon an administrative institution whose internal stability and cohesion has already been undermined by the arbitrary and piecemeal interventions of successive regimes. To be able to function effectively in such a situation, Pakistan's civil servants would have to be individuals of considerable professional acumen, integrity and initiative. But few of them today could claim to be imbued with these qualities. Given the paucity of their education and institutional environment, they are, in most cases, incapable of even comprehending the nature of the problems they face, let alone conceptualizing, formulating and evaluating the policy interventions necessary to overcome them.

3.3.2   Institutional Growth of the Military

While there has been a rapid deterioration in the level of professional competence, and in institutional procedures for decision-making and an absence of effective methods of in-service training in the bureaucracy, the military has by contrast seen a significant improvement in each of these spheres.

Unlike their peers in the civilian bureaucracy, military officers have to study, acquire new skills and pass examinations at each stage of the promotion ladder. Over the past 40 years, Pakistan's military has developed a sophisticated educational infrastructure from military public schools, through specialized colleges for professional training in various fields of engineering, electronics and aeronautics, to high-quality command and staff training institutions.

The two institutions in the latter category – the Command and Staff College Quetta (for Majors and Lieutenant Colonels) and the National Defence College Rawalpindi (for Brigadiers and above) – not only provide training in defence planning and war-gaming at the highest international level, they also enable officers to conduct interdisciplinary studies in national policy analysis in the fields of foreign policy, internal security and economic policy. The quality of the teaching staff, the methods of instruction, and the intensity and rigour of the study programmes make them into genuine centres of excellence.

One of the senior instructors at the Command and Staff College, when asked about the guiding principle of their training programme, replied: 'To develop a mind that can think on its own, that does not take anything for granted.' It seems indeed ironic that the notion of the critical mind charged by the spirit of enquiry, which over the past 40 years has been gradually banished from educational institutions in civil society, now constitutes the basis of education in the higher military institutions. Officers study long hours, use the library intensively, engage in high-quality seminar discussions and write policy papers – all activities generally absent from the civilian sphere. It is not surprising that military officers trained at such institutions develop a far more sophisticated understanding of governance than any products of civilian educational institutions in contemporary Pakistan.

Apart from the quality of intellectual training imparted to the military officers, the decision-making structure and coordination amongst the various services (army, navy, air force) have also improved. In the bureaucracy, contrary to service rules, there is political interference in promotions, appointments and operational decisions. In sharp contrast to the bureaucracy, the military has not only strengthened and professionalized its internal decision-making, but has also increasingly insulated itself from involvement of civilian authority at both administrative and operational levels, even in spheres which could be legitimately regarded as the domain of civilian executive authority. For example, the Prime Minister can make appointments, promotions and transfers up to the rank of Lieutenant General under the law. Four-star generals or service chiefs are supposed to be appointed by the President. In 1988, when General Zia-ul-Haq, the then Chief of Army Staff, sent the name of Major General Pir Dad Khan to Prime Minister Junejo for signing the order of promotion to Lieutenant General, Junejo refused, on grounds that a general who was responsible for losing Siachin did not deserve to be promoted, and, in fact, suggested to Zia that Major General Shamim Alam Khan should be promoted instead. There was a deadlock on the issue, with Zia refusing to withdraw Pir Dad Khan's name. Finally, a compromise was struck and both Major General Pir Dad Khan and Major General Shamim Alam Khan were promoted to the rank of Lieutenant General.

Another case that occurred under the public gaze involved the famous order by Prime Minister Benazir Bhutto to retire Admiral Sirohey. The officer in question had been appointed Chief of Naval Staff in 1986. Before his three-year term ended, he was appointed Chairman of the Joint Chiefs of Staff Committee (JCSC) in 1988. But in 1989, the Prime Minister decided to retire him, on the following grounds: (1) whereas the President was the appointing authority for this rank of officer under the Constitution, the Prime Minister had the authority to retire him; (2) the retirement of Admiral Sirohey fell due three years after his appointment as Admiral, i.e. in 1989. The President, supported by the military, took the contrary view: namely, that Admiral Sirohey's retirement became due not three years after his appointment as Admiral but three years after his appointment as Chairman JCSC, i.e. in 1991; and that the President was *both* the appointing and the retiring authority. This contention became public and was reported in the press.

Eventually, as a result of this pressure, the Prime Minister was obliged to let Sirohey quietly continue in office.

The Afghan operation provides another example of the military achieving institutional insulation from civilian authority even where important foreign policy considerations were involved. This operation involved providing material support to Afghan Mujahideen more or less autonomously from civilian authority, even after the latter had signed the Geneva Accord which formally committed the Pakistan government to non-interference in the internal affairs of Afghanistan.

Thus, the military has become increasingly sophisticated in terms of the quality of its professional expertise and the structure of decision-making, and has achieved greater insulation from interventions by civilian authority. At the same time, it has developed a powerful corporate image of itself. The officers owe their privilege, prestige and economic welfare to that organization. Even after they retire they know they will be looked after, with a whole range of military-run welfare societies, housing societies and manufacturing units where post-retirement service can be sought. Whereas morale and esprit de corps have risen rapidly in the army after the 1971 fiasco, the bureaucracy has undergone a gradual decline in its morale over the past three decades.

3.3.3   Relations between Military and Bureaucracy

Relations between military and bureaucracy over the past four decades have been determined partly by the differing internal processes of change in the two institutions and partly by pressures emanating from civil society, on the one hand, and the international environment on the other.

We may discern four broad phases in relations between the military and bureaucracy.

- 1951 to 1958. During this period there was an alliance between the bureaucracy and the army through the 'gang of four' consisting of Ghulam Muhammad, Chaudhry Muhammad Ali, Iskandar Mirza and General Ayub Khan. The dominance of the bureaucracy supported by the army vis-a-vis the political system can be judged from the fact that in April 1953 the then Governor General Ghulam Muhammad, who was an old bureaucrat, dismissed the Khawaja Nazimuddin government even though the Constituent Assembly had given it a vote of confidence. Soon afterwards, the Constituent Assembly met again and passed another vote of confidence, this time in favour of the new Prime Minister, Muhammad Ali Bogra, who had been nominated to this office by the Governor General. Not only did the Governor General appoint the new Prime Minister, but he also nominated ministers of the cabinet and assigned to them their respective portfolios. Thus, state power effectively passed into the hands of the Governor General and the bureaucracy and military, whose interests he pursued. The function of the Constituent Assembly was reduced merely to rubber-stamping his actions.

- 1958 to 1968. There was a formal military takeover by General Ayub Khan in 1958 (a process that had begun in 1951). Soon after the coup d'e-tat, Ayub Khan began to constitute a civilian structure of government which was formally established with the introduction of the system of 'Basic Democracy'. Under this system the President was to be elected not through direct popular vote but indirectly through an electoral college of individuals called 'Basic Democrats' (BDs) who, in turn, had been elected through elections to local bodies at the village level. Given the structure of political power at the village level, based on clans and *biradris* of the landed elite, the composition of this electoral college overwhelm-ingly favoured the interests of landlords and rich peasants. These influential landlords who were instrumental in getting the BDs elected had direct links with the bureaucrats. Thus, the BD system, in effect, con-stituted an instrument through which the bureaucracy could have an outreach into the village level clans and *biradris*, and could maintain the political system of the Ayub regime. During the Ayub regime there was a power-sharing arrangement between the Army and bureaucracy, with the bureaucracy the dominant partner. An important factor explaining why the internal balance of power within the state structure shifted into the hands of the bureaucracy after the 1958 military coup, was that both Ayub Khan and the military behind him recognized the experience and ability of the civil bureaucracy in wielding state power. Equally important was the fact that the bureaucracy at that stage could still boast of highly competent professional administrators inherited from the ICS tradition, and an institutional cohesiveness in its decision-making structure.
- 1971 to 1977. During the early period of the military regime of General Yahya Khan (1969–71) the bureaucracy had been relegated to a relatively minor role compared to the military, in the task of governance. The bureaucracy had also been fragmented and demoralized by the dismissal of 303 civil servants during the regime of General Yahya Khan. The sub-sequent period under Prime Minister Zulfiqar Ali Bhutto saw the further fragmentation and demoralization of the bureaucracy. The new Bhutto government carved out from the bureaucracy a personalized chain of command through the appointment of politically loyal individuals in key positions. At the same time, an attempt was made to reduce the power of the elite CSP (Civil Service of Pakistan) cadre of the bureaucracy. This was done first by purging 1,300 officers on grounds of misuse of power, and filling their vacancies with individuals personally loyal to Bhutto. These were drawn either from other sections of the civil administration or from outside the bureaucracy, by instituting a system of 'lateral entry', as mentioned under section 3.3.1 above. By thus short-circuiting the hier-archy of the CSP and penetrating it with the officers loyal to the PPP, large sections of the bureaucracy were politicized and made amenable for direct use by political forces.
- 1977 to 1988. During this period President General Zia-ul-Haq stabilized and rehabilitated the bureaucracy, although it was very much a junior

partner to the military in the task of governance. He created a clear demarcation of roles. The military formulated the policy and the bureaucracy was made responsible for implementing it. Although the General relied on the military for his power, even in the daily running of state affairs – there was a regular meeting of the Corp Commanders and Principal Staff Officers under the Chairmanship of General Zia-ul-Haq in his capacity as Chief of Army Staff, to discuss national policy – the General also maintained three senior bureaucrats as close confidants in the administration. They were Secretary General Ghulam Ishaq Khan, Interior Secretary Roedad Khan and Defence Secretary Ijlal Haider Zaidi. Up until his retirement in 1982, Agha Shahi was also an influential bureaucrat on whom the General relied to implement the foreign policy of what was essentially a military regime.

The history of the changing balance of power between the army and bureaucracy in Pakistan shows a rapid increase in the weight of the military relative to the bureaucracy in determining national policy in foreign policy, economy and internal security. This shift was due not merely to the weakening of civil society relative to the state apparatus as a whole but, equally importantly, to the institutional deterioration of the bureaucracy as an arm of governance.

3.4 The Structure of State Power and the People of Pakistan

At Independence in 1947, the bureaucracy and the army held a predominant position in the state power structure relative to the institutions of civil society. This was due first to the form of the freedom struggle on the one hand, and the nature of the Muslim League on the other. Since the freedom struggle was essentially a constitutional one, the state apparatus of the colonial regime remained intact, albeit in a weakened condition. The bureaucracy, which constituted the steel frame of the Raj and the army, continued after the emergence of Pakistan to determine the parameters within which political and economic changes were to occur. However, as noted, the position of the military relative to the bureaucracy within the power structure became increasingly important, partly because of the different internal dynamics within each of these two institutions.

The second factor in the failure to subordinate the state apparatus to the political system lay in the two basic characteristics of both the Muslim League before partition and the PPP during the two decades between 1970 and 1990.

- In the pre-Independence period both the Muslim League as well as the Pakistan People's Party were movements rather than parties. They were therefore unable to establish an organizational structure or develop a political culture on the basis of which people's power of the people could be institutionalized and used to subordinate the army and the bureaucracy to a stable political system.

- The Muslim League in the decade before partition, and the PPP during the early 1970s, were taken over by landlords whose political interest lay in constraining the process of political development and, while ruling in the name of the people, in confining politics to a struggle for sharing the economic spoils amongst various factions of the political elite.

The political elite in Pakistan has so far demonstrably failed to build within the state of Pakistan a modern democratic polity marked with social justice, as envisaged by the founding father, Quaid-e-Azam Muhammad Ali Jinnah. This would have meant building institutions through which the will of the people could become operative within the power structure, developing a political culture which could strengthen and sustain these institutions, and finally, initiating an industrialization process through which the people of Pakistan could make a contribution to the contemporary world. Members of Pakistan's political elite have generally preferred narrow personal gain to national interest, and have engaged in internecine quarrels fuelled with greed in situations which required unity and self-sacrifice for the nation.

Yet, despite the failure of the political elite, the dominance of the military in the structure of state power and growing social polarization, it is remarkable that whenever the people as a whole have intervened, not only have they shown a high level of political consciousness but, it can be argued, their political maturity has grown over time. For example, in 1956 when Western powers were involved in a conflict with Nasser's Egypt, even though the government and the political elite supported the Western allied powers, the people of Pakistan came out on the streets in large numbers to voice their support for the nationalist struggle of the people of Egypt. Again in 1968, the people of Pakistan came out on the streets to express their opposition to the regime of Ayub Khan which at the political level had repressed popular aspirations, at the economic level had generated acute inequality between social groups and regions, and at the foreign policy level had compromised Pakistan's national pride in the Tashkent Agreement. After the Pakistan Movement, whose struggle for Pakistan resulted in the creation of a new state, the movement against the Ayub regime was the second great movement. It generated demands for social equality, justice and political representation of the dispossessed.

It was Zulfiqar Ali Bhutto who articulated the deep-rooted aspirations of the people during this period: in a short time-span, he was catapulted into power in 1971. Yet, within six years the people had grasped the failure of Prime Minister Bhutto to build a state structure in which power could actually go to citizens at the grass roots; a political system within which the ruling People's Party could generate new leadership at several levels of society, and an economic system under which drastic measures could be taken to alleviate poverty, unemployment, hunger and disease. The disillusionment of the people with their beloved leader was expressed by their silence when the PNA led an urban revolt to destabilize the regime of Prime Minister Bhutto. However, the enduring contribution of Z.A. Bhutto in articulating the aspirations of the poor and in giving a new dignity and pride to the wretched of the earth was acknowledged by the people of Pakistan in the widespread anguish expressed after his 'judicial' assassination. When his daughter Benazir Bhutto took on the mantle of leadership in the struggle against the dictatorship of General Zia-ul-Haq, the people once again responded with both passion and heroism. The popular struggle against the dictatorship of the General culminated first in the 1983 movement and later in the unprecedented demonstration in Lahore on the arrival of Benazir Bhutto in August 1986. But then, within 20 months after the popularly elected Prime Minister Benazir Bhutto had come into power, when the people once again went to the polls they expressed their dissatisfaction with the performance of her regime by voting in favour of the Islami Jamhoori Ittehad (IJI: Islamic Democratic Alliance), the multi-party political alliance formed against Ms Bhutto during the 1988 general elections.

Thus it is that the people of Pakistan, the poor and downtrodden, despite the erosion of institutions of civil society, have nevertheless demonstrated a high level of political consciousness and emerged as a factor to be reckoned with. It is for this reason that the military, even when there was no apparent obstacle to the reimposition of military rule, after the death of Zia on 17 August 1988, sought a civilian dispensation within which it could exercise its power as a major actor, and through which the latent tensions of the populace could be defused.

Notes

1. For an analysis of the economic strategy practised during the Ayub period see Griffin, 1974.
2. For detailed evidence on industrial concentration see White, 1972.
3. For an examination of the polarization phenomenon in Pakistan's rural sector see Hussain, 1976, 1980.
4. For detailed analysis of disparities among regions of West Pakistan see Hamid & Hussain, 1974.
5. For evidence on shortage of basic services, see Hussain, 1988.
6. For evidence on the state of Pakistan's environment see Sayyed Engineers, 1991.

References

Alavi, Hamza, 1983. 'Class and State in Pakistan', pp. 40–93 in H. Gardezi & J. Rashid, eds, *The Unstable State*. Lahore: Vanguard.

Griffin, Keith B., 1974. 'Financing Development Plans in Pakistan', pp. 31–64 in K.B. Griffin & A.R. Khan, *Growth and Inequality in Pakistan*. London: Macmillan.

Hamid, Naved, 1974. 'The Burden of Capitalist Growth: a Study of Real Wages in Pakistan', *Pakistan Economic and Social Review*, vol. 12, no. 2, Spring.

Hamid, Naved & Akmal Hussain, 1974. 'Regional Inequalities and Capitalist Development', *Pakistan Economic and Social Review*, vol. 12, no. 4, Autumn.

Hussain, Akmal, 1976. 'Technical Change and Social Polarization in Rural Punjab', pp. 316–371 in K. Ali, ed. *Political Economy of Rural Development*. Lahore: Vanguard.

Hussain, Akmal, 1980. 'Changes in the Agrarian Structure of Pakistan, with Special Reference to the Punjab Province 1960–1978'. DPhil Thesis, University of Sussex.

Hussain, Akmal, 1985. 'A Note on Rural Poverty and Agrarian Structure in Pakistan'. Paper presented at the 18th World Conference, SID, Rome, 10–14 July.

Hussain, Akmal, 1988. *Strategic Issues in Pakistan's Economic Policy*. Lahore: Progressive Publishers.

Hussain, Akmal, 1990a. 'The Crisis of State Power in Pakistan: Militarization and Dependence', pp. 199–236 in P. Wignaraja and A. Hussain, eds, *Challenge in South Asia, Development, Democracy and Regional Cooperation*. Karachi: Oxford University Press.

Hussain, Akmal, 1990b. 'Past Mistakes, Present Follies', *Newsline* (Karachi), December.

Jalal, Ayesha, 1990. *The State of Martial Law*. Cambridge: Cambridge University Press.

Qutub, Ayub, 1991. 'Walking Lightly', pp. 345–67 in A. Qutub, ed. *Towards a National Conservation Strategy for Pakistan*. Karachi: IUCN/GOP.

Sayyed Engineers, 1991. *Pakistan: Environment Under Threat, Calendar Study*. Lahore: Sayyed Engineers.

White, L.J., 1972. *Industrial Concentration and Economic Power in Pakistan*. Princeton, NJ: Princeton University Press.

4

Pakistan: the Politics of 'Fundamentalism'

ABBAS RASHID

4.1 Introduction

Essentially, the term 'fundamentalism' suggests going back to the basic texts and reproducing as closely as possible the laws and institutions found there. It has also come to imply a dogmatic adherence to tradition, orthodoxy, inflexibility and a rejection of modern society, intellectual innovations, and attempts to recreate a 'golden' era. With few exceptions, the term has been used in Western discourse to designate the moribund and hostile 'other'. What such an assessment – frequently made about Third World Muslim societies – lacks is an appreciation of the dynamic populist movement that can engender such a political phenomenon (religious piety is usually secondary), as in the case of Iran.[1] Nor is there sufficient awareness of the lack of credibility in most of these societies of 'secular' alternatives which, historically, have often failed.

To some extent, the perception of fundamentalism (or revivalist movements) in the West may have been influenced by government perceptions of the dictates of the national interest. The US government view of a threatening, or unpredictable, Iran on the one hand, allied with Saudi Arabia on the other, certainly affects how Saudi Islam is viewed in the USA for instance, as compared to the Iranian variety of Islam. In terms of orthodoxy, inflexibility and the absence of any kind of a democratic framework, Saudi Arabia as a society would certainly seem to be the more oppressive and 'threatening' – at least towards its own people.

Fundamentalism then, apart from its more ominous aspects, also suggests bypassing the clergy and going back to the original sources for guidance. In the confusion engendered by modernization, it means authenticity and rootedness which gives it strength and resilience. That all of this is not enough for it to grapple with contemporary reality, to respond to that challenge or even to release the full potential of its adherents is another matter altogether. Its moving force is still powerful and it can be extensive in its sweep. As W.C. Smith (1957, p. 156) commented about the Muslim Brotherhood Movement:

> It represents, in part, a determination to sweep aside the degeneration into which Arab society has fallen, the essentially unprincipled social opportunism interlaced

with individual corruption; to get back to a basis for society of accepted moral standards and integrated vision, and to go forward to a programme of active
implementation of popular goals by an effectively organized corps of disciplined
and devoted idealists.

This may have been overstating the case. Nevertheless it admirably draws
attention to aspects that, for instance, enabled the Khomeini-led 'fundamentalists' in Iran to acquire such enormous popular support. The
anti-Westernism so evident in the Iranian movement is a feature usually central to the contemporary fundamentalist resurgence throughout the Muslim
world (Rahman, 1982).

4.2   Revivalism and Reconstruction

This anti-Westernism was, however, not the main thrust of the Muslim
revivalist movements over the past two centuries. Beginning with the Wahabi
movement led by Ibn-Abd-al-Wahab in the 18th century in Arabia, the
emphasis was on internal reform. While a similar dynamic for reform and
regeneration within Islam was working itself out during the 19th century in
Africa and the Indian sub-continent, another significant movement of
'Islamic modernism' arose in Turkey, Egypt and India, countries influenced
by their encounter with Western ideas and power (Rahman, 1979).

The Wahabi movement in Saudi Arabia and related ones such as the
Sanusis in North Africa, and the Islamic reform movements in India had certain common characteristics. One was the impetus to transform Muslim
society, given its state of socio-moral degeneration, and to do this by going
back to original Islam. This was to be done by challenging the finality and the
rigidity of the traditional schools of law through *Ijtehad*; i.e. rethinking the
meaning of the original message (Quran) by the learned in Islam; and by
doing away with what were seen as the debilitating effects of corrupted Sufi
practices that had promoted laxity and dissipation in Muslim society
(Rahman, 1979, p. 317).

In India, Shah Waliullah (1703–81), an intellectual with a Sufi orientation,
sought to integrate in his work the essentials of the *Sunnah* (practice of the
Prophet Mohammad) with a purified Sufism by bringing together *Hadith*,
Fiqh, theology, philosophy and Sufism (Rahman, 1979, p. 318).

Subsequently Jamaluddin Afghani extended this concern to combine external reform with internal defence (Smith, 1957, p. 47). He perceived clearly not
only the internal inadequacies but also the external: a feeble Muslim world
threatened by a powerful West. Hence in addition to his efforts at reform he
emphasized the need for reconciliation among the Shia and Sunni. He was
directly involved in agitational activities against British imperialism. His influence reached from India to Iran, the Arab world and Turkey. Significantly, he
may have been 'the first Muslim revivalist to use the concepts "Islam" and the
"West" as connoting correlative – and of course antagonistic – historical
phenomena' (Smith, 1957, p. 49).

Another common factor in fundamentalism is that of selectivity. Even when there is a genuine desire to go back to the source, circumstances and people's felt needs usually dictate otherwise. The Wahabi movement, for example, saw the need for ridding Islam of Sufi superstition and found its justification in scriptural sources. Yet neither sought nor rediscovered the basic message of the Quran barring the economic exploitation of man by man. This also escaped the latter-day movement of Islamic modernism (Rahman, 1979, p. 319).

In one sense, pre-modernist revivalism, as it has been called, was a liberating force in that it reaffirmed the right of *Ijtehad* (Rahman, 1979, p. 319). And yet it could make no significant contribution to a renaissance of Islam. What other reasons there may have been for this lack of readiness on the part of Muslim societies to make the leap from theory to practice, one crucial aspect had to do with the utter bankruptcy of the traditional educational system of Islam, which had become devoid of originality and encouraged conservatism and caution rather than challenge and enquiry. It had, nevertheless, a rich and sophisticated tradition with which to work. Had the revivalists dispensed with it in favour of a dynamic intellectual framework and an encompassing world-view based on the Quran and *Sunnah*, the results might well have been very different. Instead, they more or less limited themselves to the Quran and *Hadith* without developing any new methodology as to how these were to be taught and understood (Rahman, 1979, p. 319). Paring down the *madrassa* (religious school/seminary) syllabus, then, to the fundamentals meant not so much the elimination of the corrupting influences of the intervening ages and the confusion thus engendered, but being 'limited' to a knowledge of the fundamentals of the faith without understanding them in the context of contemporary reality.

The revivalist effort was constrained by a traditionalist intellectual framework, also on a broader sociopolitical level in India, in the efforts of the *ulema* (religious scholars) of Deoband. Next to Al Azhar of Cairo, this was the most important theological academy of the Muslim world. Influenced by the Wahabi and the Waliullahi movements, some of its theologians, such as Ubaidullah Sindhi, actively participated in the resistance to British imperialism. And yet Deoband kept the doors of *Ijtehad* firmly closed and its emphasis on rationalism confined within an orthodox intellectual framework. Its concern, therefore, with changing the material condition of Indian Muslims did not have major results (Smith, 1969, p. 363). And this was equally true of the other important seminaries that sought to cope with the deterioration within, as well as the challenge of British imperialism from without. The differences in their approach are cogently summed up by W.C. Smith (1969, p. 362):

> We shall notice the Bareilly school as accepting without criticism the social and religious condition of the masses and of the old order in all its collapse; the important Deoband academy as accepting the old order in principle but trying to revive and purify it; and the Farangi Mahal and Nadwat ul Ulema in Lucknow as representing a partial and quite unsuccessful attempt to incorporate something of the new order into the old Islam.

In India, Islamic modernism – or the reconstructionist enterprise – got firmly under way after the war of independence of 1857 with the efforts of Syed Ahmad Khan, who stood for a liberal and rationalist Quranic interpretation and social reform, particularly in the field of education. Sir Syed was primarily a social reformer who sensed a need to address the issue of religion. The Hindus were 'exceedingly indifferent about speculative doctrine', he wrote when explaining to the British the Muslim role in the 1857 war of independence, or Mutiny as it was referred to. He said (Khan, 1873, p. 23):

> The Muhammadans, on the contrary, looking upon the tenets of their creed as necessary for their Salvation and upon the neglect of them as damnation, are thoroughly well grounded in them. They regard their religious precepts as Ordinances of God.

Going back to the original sources, Syed Ahmad held the Quran to be determinative of our understanding of Islam. Taking Waliullah's brand of fundamentalism as his point of departure, he sought a rationalist exposition of Islam that represented a qualitative change from the past into the modern era. To this end he invoked such anti-traditionalist sources within Islamic history as the views of the Mutazzila and of the *Ikhwan al Safa* (Brethren of Purity) (Ahmad, 1967, p. 41). Starting afresh with the Quran and bringing out its relevance to the new society of his day, he rejected the canonical traditions and the authority of the four accepted legal schools. In interpreting the Quran he focused squarely on its principles (*usul*), asserting that the details of specific historical situations were not significant (1967, p. 42).

Sir Syed emphasized *Ijtehad* as the right of every individual Muslim and rejected *Ijma* (consensus) in the classical sense of the *ulema* (1967, p. 54). Though Sir Syed's approach to Islam and the Quran did not gain currency on a popular level (just as that of the *ulema* did not), he did lay the groundwork for people like Ameer Ali, Chiragh Ali, Iqbal and others who came later to interpret Islam from a liberal standpoint. His success was more immediate and visible in the field of education. The Mohammadan Anglo-Oriental College which he founded at Aligarh had a profound impact on the Muslim middle class of North India, an impact which was to play a crucial role in the politics of Indian Muslims subsequently.

Mohammad Iqbal, poet, philosopher and political thinker, rejected, in the tradition of Sir Syed, the static traditionalist interpretation of Islam, asserting that the Quran provides an essentially dynamic world-view for Muslims. He pointed to *Ijtehad* as proof of this assertion (Iqbal, 1960, pp. 147–148):

> . . . eternal principles when they are understood to exclude all possibilities of change which, according to the Quran is one of the greatest signs of God, tend to immobilize what is essentially mobile in its nature . . . What then is the principle of movement in the nature of Islam? This is known as *Ijtehad*.

As a source of law, he argued, *Ijtehad* could not remain the prerogative of individual representatives of the classical schools of law. Nor could *Ijma* alone suffice (Ahmad, 1967, p. 155). Both functions, he maintained, could now be usefully performed at the level of a Muslim Legislative Assembly where *Ijtehad* by elected representative (those having the confidence of the community) could then take the form of *Ijma* among the various sects.

While Iqbal conceded the need for some representation in the Assembly for the *ulema* (Ahmad, 1967, p. 155) his emphasis on representative democracy within the framework of Islam was clearly evident. From a preoccupation with pan-Islamism, Iqbal slowly gravitated towards the idea of a separate homeland for the Indian Muslims. He rejected the idea of a polity on Indian national lines 'in which the religious attitude is not permitted to play any part . . . if it means the displacement of the Islamic principles of solidarity' (1967, p. 160).

The creation of Pakistan was not the work of clerics and religious divines: it was the result of the efforts of a liberal Westernized leadership that successfully articulated the aspiration of the Muslims in different parts of India for substantially improved material conditions and an absence of Hindu economic and cultural domination. The motive force for Pakistan came largely from the middle class Muslims of North India, many of whom had been educated at Aligarh. At some level, they identified with the ideas of Syed Ahmad Khan and Mohammad Iqbal. The leaders of the Muslim League – including Mohammad Ali Jinnah – the founding father, may have thought sincerely about the application of Islamic principles, but they certainly did not regard the movement for Pakistan as an effort to recreate some kind of a 'golden age' in Islam or to re-establish Wahabiism. Nor did the large number of Muslims who opted in favour of the somewhat 'secular' leadership of the Muslim League (rather than the religious leadership) thereby reject Islam, for the *ulema* do not derive their authority from within the Quran and *Sunnah*. There is in Islam no Church or priestly class, counterposed to the state and enjoying a kind of divine sanction. In the minds of most, whatever the other inducements underlying the demand for Pakistan, the project certainly had an idealist element involving the issue of identity and Islam.

Thus, in 1947 the Muslim League leadership found itself with a nation-state on its hands, and a people who sought to order their lives according to Islam in a broad sense but with very little agreement on what Islam required of them. For the great majority this was not a pressing issue. Yet it was a situation that affected the *ulema*, who had, with very few exceptions,[2] opposed the founding of Pakistan but had migrated in great numbers from India, to establish themselves in the newly formed state.

From the start they challenged the leadership on the issue of the Islamic nature of the Pakistani state. Even Jinnah (quoted in Abbot, 1968, p. 188), while emphasizing the broad principles set forth by Islam, had to concede its centrality:

Islam is not only a set of rituals, traditions and spiritual doctrines, Islam is a code for every Muslim which regulates his life and his conduct in all aspects, social, political, economic, etc. It is based on the highest principles of honour, integrity, fairplay and justice for all. One God, equality and unity are the fundamental principles of Islam.

The social approach to Islam, as distinct from religion (and religiosity), was even more emphatically stated in his speech to the Constituent Assembly:

> You may belong to any religion, caste or creed – that has nothing to do with the business of the state. . . . We are starting with this fundamental principle that we are all citizens and equal citizens of one State . . . Now I think you should keep that in front of us as our ideal, and you will find that in course of time Hindus would cease to be Hindus and Muslims would cease to be Muslims, not in the religious sense, because that is the personal faith of each individual but in the political sense as citizens of the state. (Quoted in Mortimer, 1982, p. 20)

From this position in the first heady days of Independence in August 1947, the League leadership allowed the debate on the kind of state Pakistan was going to be to shift significantly in favour of the 'Islamist' lobby. It introduced in the Constituent Assembly a resolution defining the 'Objectives' of the new state only two years later in 1949. Subsequently, with some modifications, this became the preamble to the Constitution and eventually was made a substantive part of the Constitution.

The Resolution affirmed that 'sovereignty over the entire universe belongs to God Almighty alone', that therefore the people of Pakistan were to exercise power only 'within the limits prescribed by Him' and that Muslims would be enabled to order their lives 'in the individual and the collective spheres in accord with the teachings and requirements of Islam as set out in the Holy Quran and *Sunnah*' (Ahmad, 1967, p. 238). While the resolution reaffirmed democratic principles, the very fact that it was passed bolstered the Islamists' claim to being a significant component in the new setup. Indeed Maulana Abul Ala Maududi's Jamaat-e-Islami claimed that the Resolution was a result of their efforts.[3]

Maududi's narrow, though logically coherent, interpretation of the *Shariah* (Islamic law) and his theory of the Islamic state held a certain kind of appeal for a section of middle-class and lower middle-class Muslims. Underlying his elaboration of what an Islamic state should be was the premise that Islam was engaged in a broad conflict with the West, and that modernization, as dictated by the West, meant also a surrender of Muslim identity and culture. Hence genuine improvement of the Muslim lot meant less, rather than more, Westernization which had become, erroneously, a synonym for modernity. As against the modernizers, then, Maududi sought to extend the areas of activity governed by a strict code of Islamic rules and laws, and to transform Islam from a faith into a system (Smith, 1957, p. 215).

The central idea in Maududi's formulation is sovereignty, in the sense of absolute power and authority, which belongs to God. In a political context the implication is that those capable of understanding and interpreting the Quran and the *Shariah* should command the loyalty and obedience of the people in an Islamic state:

> As the West understands it democracy encompasses the concept of sovereignty (*Hakimiyya*). And what we Muslims call democracy refers only to *Khilafat* (vice-regency). In their democracy, to run the state, the government is formed and changed

by popular vote. So, too, in our democracy. The difference is that in their conception the democratic state enjoys sovereignty (absolute) whereas a democratic *Khilafat* is bound by the Law of God. (Maududi, 1981, p. 320)

Elsewhere, elaborating on these ideas, he has proposed that the elected head of state should be the supreme head of legislature, executive and judiciary, alike. Further, opposition in the form of political parties and cliques in the legislative assemblies should not be allowed. In his view, non-Muslims are not eligible for key posts and should be allowed to exercise the right to vote only under a system of separate electorates. Women could not become head of state.

Most of these ideas, modified at times under pressure of circumstances, were adopted by his party —the Jamaat-e-Islami. However, apart from his puritanical creed and a lukewarm, somewhat suspect commitment to democratic norms, the JI's limited popular support also owed something to the proto-Fascist tactics employed by it, particularly on campuses, and to the contradiction between its high-sounding moral postulations and actual political practice.

4.3 The Secular Response to 'Islamist' Politics

The leadership that had assumed control of the new state, though committed in broad terms to Islam, was largely Westernized and secular in outlook. It represented the emerging Muslim bourgeoisie and the feudal elite (along with the salariat) which had internalized the ideals and the idiom of Western secularism. They held religion to be a largely private matter between man and God. In all the branches of government, the armed forces, industry, education, etc., such men were dominant.

Initially the Westernized leadership treated efforts by the Jamaat-e-Islami and other Islamist parties with their base among the *petit bourgeoisie* and sections of the middle class as something of a nuisance to whom some concessions had to be made from time to time. Along with the intelligentsia they believed that it was only a matter of time before the forces of modernity and secular liberalism would completely marginalize such elements in society. They failed to recognize that:

> Liberal secularism is itself a faith, a positive conviction. It has its own foundations, moral and intellectual; its own martyrs and heroes and ideals; its own history; and its own institutions. Some expect it to appear of itself so soon as religious faith is circumscribed or dropped. This is glib. (Smith, 1957, p. 208)

In addition, the Western liberal idea in Pakistan was constantly undermined by the fact that those who professed to support it were content to leave unchanged a system that incorporated the worst excesses of feudalism and an unjust social order.

Less than six years after Pakistan's inception, the Islamists were able to put up a remarkable show of strength and expose the weakness of the government and the modernist liberals. Starting in March 1953, almost for a month and a half, widespread disturbances in the Punjab resulted from the government's rejection of their demand to declare the Qadiani Ahmadi sect as a non-Muslim minority and to remove from office the foreign minister of Pakistan along with other Ahmadis occupying key posts within a month. Failing this, the Action Committee constituted by the All Pakistan Muslim Parties Convention held in Karachi threatened direct action. The rejection of the ultimatum, followed by the arrest of the *ulema* involved, led to a series of public meetings and demonstrations. There were incidents of shooting, looting and murder at various places. Yet as the report of the judicial commission instituted to investigate the incident states:

> In the meeting of citizens at Government House on the afternoon of the 5th of March no leader, politician or citizen was willing to incur the risk of becoming unpopular or marked by signing an appeal to the good sense of the citizen. (Government of Pakistan, 1954, p. 234)

Eventually, the military had to be called in. Martial Law, proclaimed in Lahore, remained in force until mid-May.

The 1953 disturbances and the reaction to them highlighted several important points. First, at the level of civil society there was no response from the powerful section of Westernized liberals, including those many who saw Islam not as just a set of rituals but who emphasized its broad principles and spoke for modernity and the bourgeoisie. Secondly, the Punjabi elite, together with its Urdu-speaking counterpart, hit upon the idea of using Islam as a counterpoint to the rising trend of regional and ethnic identification and demands of the Bengalis, Sindhis, Pashtun and Baluch (Alavi, 1986, p. 29). Among fellow Muslims, quibbling for a greater share of economic resources or political power could be dismissed as irrelevant or worse, subversive in the context of the (almost) sublime project! Thirdly, the Islamist parties, including the Jamaat-e-Islami which at the time had a membership only of 999 persons (*Government of Pakistan*, 1954, p. 243) became keenly aware of the potential for agitational politics under the cover of sensitive religious issues. The dislocation they could cause by playing upon the emotions of the more fanatical or easily swayed members of society meant gaining a bargaining position that they could not have dreamed of achieving solely on the basis of the number of followers.

Successive civilian governments sought the short-cut of adopting an Islamic formalism to assuage what they saw as potentially disruptive forces in society. For instance in the 1956 Constitution, the 'directive principles of state policy' provided that the 'state shall endeavour . . . to make the teaching of the Holy Quran compulsory (for Muslims); to promote the unity and observance of Islamic moral standards; and to secure the proper organization of *zakat* (charitable tax), *waqfs* (religious endowments) and mosques'. Also the state was to 'endeavour' to 'prevent prostitution, gambling, the taking of drugs and the consumption of alcoholic liquor other than for medicinal purposes' (Baxter et al., 1988, p. 173). At the same time efforts were made to provide a liberal interpretation of Islamic injunctions. In 1955, for instance, a seven-member commission was appointed to study the existing laws of marriage, divorce and family maintenance to see if they could be modified to give women their proper status as prescribed by Islam. The commission's report recommended liberal reforms in the existing laws, but its recommendations were ignored. They had to await the military government of Sandhurst-trained General Ayub Khan for implementation, in a limited form, under the Muslim Family Laws Ordinance of 1961.

There was also an attempt at a more institutionalized response to counter conservative religious influences, particularly that of Maulana Maududi and the Jamaat-e-Islami. For instance, the Institute of Islamic Culture was set up in 1954 – a year after the anti-Ahmadi agitation. It was headed by Khalifa Abdul Hakim, who maintained that: 'Islam is not the name of any static mode or pattern of life; it is spirit and not body; it is aspiration and not any temporal or rigid fulfillment.' Islam, he said, now lies buried under 'heaps of retrograde legalism and life thwarting practices' over centuries during which the West progressed and began to describe Islam as an 'outworn creed incapable of adaptation to changing circumstances' (Abbot, 1968, p. 208). Reasonable as all this may seem, its impact at the popular level was very limited – not least because of the overt and seemingly obsequious deference to Western norms and judgement. It was the kind of opening that Islamists of various hues had exploited against Syed Ahmad Khan, Amir Ali and others who had held up the West as a kind of model and shown themselves to be particularly concerned about what the West thought of them. Khalifa Abdul Hakim elicited a similar response from his detractors.

Work on a liberal Islamic interpretation was also carried on under Ayub Khan's regime by Fazal-ur-Rahman, director of the Islamic Research Institute (1962–68). But as he wrote (1982, p. 124), he was unable to bridge the gap between 'tradition' and 'modernity':

> The case of this institute illustrates the real dilemma of the purposeful and creative Islamic scholarship. On the one hand are the traditional *Madrassas*, which are incapable of even conceiving what scientific scholarship is like and what its criteria are. On the other hand there has been a constant flow of those scholars who have earned their PhDs from Western universities – but in the process have become 'orientalists'. That is to say, they know enough of what sound scholarship is like, but their work is not Islamically purposeful or creative. They might write good enough works on Islamic history, or literature, philosophy, or art, but to think Islamically or to rethink Islam has not been one of their concerns.

He might have added that usually the liberal concern with such an effort does not go very deep. Under pressure, the liberals have quickly yielded to the Islamist lobby for reasons of expediency and convenience – as in the case of Fazal-ur-Rahman himself, who was eventually forced to resign from the Institute which then, effectively, collapsed. The bourgoisie and liberal intelligentsia felt no need to take a stand in his support. Possibly the Institute's identification with the Ayub government was a factor in its isolation. In any case, one unfortunate result was that Fazal-ur-Rahman had to leave the country.

More often than not, such efforts lacked credibility, stemming from an opportunistic and often cynical use of religion by those who were seen as liberal and secular in outlook. The 'modernist' Ayub Khan, credited with implementing the Family Laws Ordinance on the basis of proposals made by a commission that sought to improve the status of women in the country, was not above recruiting the *ulema*'s help for his election campaign against Jinnah's sister Fatima Jinnah. He got together some obliging *ulema* to issue a *fatwa* (religious edict) on the eve of the elections, declaring that a woman could not be head of state in an Islamic country.

Similarly, Zulfiqar Ali Bhutto – with his powerful slogan of 'bread, clothing and shelter' and who pushed aside the Islamist parties in the 1970 elections – only four years later presided over the decision of his party in the National Assembly to declare the Ahmadiya community a non-Muslim minority. That was a decision that far more conservative political regimes had resisted in the face of much greater pressure. Before long, towards the end of his rule, Bhutto tried 'Islamist' tactics in a crude, mechanical fashion. In a desperate bid to stay in power he declared Friday, instead of Sunday, to be the weekly national holiday and banned the consumption of alcohol.

It can be argued that had Bhutto managed to deliver on his radical promises or had he, at least, refrained from increasing authoritarianism, then the urban middle class and Islamically articulated movement *Nizam-e-Mustapha* (system of the Prophet's time) against him would have been less effective. A more impressive show of support from a grateful and mobilized populace might also have given the army second thoughts about a takeover. But then it is precisely this inability to deliver on the part of those who claim the mantle of modernity or radicalism that, by default, strengthens the 'Islamist' enterprise. This is not to say that the Islamist parties have not shown themselves to be opportunist and do not suffer from a similar lack of credibility. The JI, as mentioned earlier, supported Fatima Jinnah against Ayub Khan, even though it had earlier held that a woman could not be head of state. Again, much to popular disgust, the JI was quick to embrace the army dictatorship of General Zia-ul-Haq. More generally, the commitment of Islamist parties to socio-economic change is also perceived as lukewarm. Three factors have helped them along in the more recent past: their emphatically projected anti-Westernism, enhanced financing from external sources and greater support and cooperation from the state.

4.4 A Major Step Forward

The assumption of power by General Zia-ul-Haq following a military coup in July 1977 represented a major step forward for the politico-religious forces and a qualitative change from the preceding state of affairs. So far the dominant forces of society and state had viewed the religious establishment with considerable suspicion, using them when convenient, giving in to their demands when expedient, perhaps to avoid what could prove to be costly confrontation. General Zia-ul-Haq, risen from the ranks of the *petit bourgeoisie*, like many others of his class, saw a certain kind of Islamic framework as the fulfilment of Pakistan's destiny as well as the answer to his regime's struggle for legitimacy.

After roughly a year and a half of neutralizing his regime's political opponents, Zia announced his intention to introduce an Islamic system in the country. He started with a series of 'reforms' designed to bring laws in various areas of activity in conformity with the tenets of Islam. Many of the changes had the imprint of the Jamaat-e-Islami, from which members were soon provided for the Zia cabinet. Support for Zia's 'Islamic' package was formidable, especially in the Punjab. It came from the lower middle class in the urban centres, part of the bourgeoisie that had been disenchanted with Bhutto's style of governance and policies (such as nationalization of some major industries), as well as artisans, petty shop keepers and workers returning from the Middle East and elsewhere. The latter had improved their economic standing but had remained socially and politically conservative. They felt more comfortable with Zia's Islamic, status quo, orientation as opposed to Bhutto's socialist rhetoric, Sindhi identity and his strong personal style.

Clearly Zia's own priority was to give himself unfettered power. The Provincial Constitutional Order which Zia promulgated as Chief Martial Law Administrator in 1981 contained a number of provisions which unambiguously concentrated power in his hands. For instance, Article 4, dealing with the Federal Council, stated that:

1   There shall be a Federal Council (*Majlis-e-Shura*) consisting of such persons as the President may, by Order, determine.
2   The Federal Council (*Majlis-e-Shura*) shall perform such functions as may be specified in an Order made by the President (from Mortimer, 1982, p. 222).

Article 9 restricted the power of the High Court, among other things, with respect to convictions by a military court or tribunal established under martial law orders or regulations. Under the provisions of Article 14, with the exception of the Jamaat-e-Islami, all political parties were required to have the written permission of the Chief Election Commissioner (appointed by the President) in order to function. Further, all judges were required to take a new oath of office swearing to abide by this 'Constitution Order', and a number of senior judges were dismissed for refusing to do so (Mortimer, 1982, p. 223).

In substance there was little in all such measures to distinguish Zia-ul-Haq and his regime from any other dictator or martial law regime. An elaborate system of martial law courts and tribunals was put in place and the power of the higher judiciary drastically curtailed.

The more high-profile 'Islamic' changes brought about by Zia were meant to provide legitimacy for his regime as well as strengthen his constituency among the *petit bourgeoisie* as well as that section of the middle class and the feudals who favoured a strong centre. The latter mostly managed to stay above the law in any case and were therefore not particularly concerned about such things as the mediaeval system of punishments being instituted.

Under the Hadood Ordinances – a collection of criminal laws promulgated by President Zia in 1979 and enforced in 1980 – adultery became punishable by stoning to death, theft by amputation of the right hand and left foot, the drinking of alcohol and perjury (false accusation of adultery) with 80 lashes. However, there were numerous instances of flogging instead of executions sanctioned by provisions of martial law, since the rules of evidence under Islam are far too stringent to make conviction a simple matter (Mortimer, 1982, p. 225). There were no cases of amputation or stoning carried out under this law. The Pakistan Medical Association declared that its members would not carry out any such surgery. Nevertheless, such laws, along with public hangings and lashings, created an atmosphere that encouraged excesses in the name of Islam and a kind of vigilante mentality. The stoning of a child, denounced as illegitimate, outside a mosque in Karachi was widely reported; women were paraded around naked to 'avenge family honour' in the Punjab; the local *maulvi* (priest) and his minions were reported as making veiled threats to those in the neighbourhood not seen in the mosque regularly, etc. That the intent was to terrorize political opponents as well as potentially troublesome sections of society (such as lawyers and journalists, some of whom were jailed and whipped during the Zia regime) rather than to Islamize Pakistan became clear fairly early on. It is interesting, however, that punishments such as hand amputations and public lashings had been announced, prior to the Islamization process, as part of the martial law setting (Noman, 1990, p. 122).

The power of the judiciary was curtailed in various ways. For instance, superior courts could not challenge executive decisions emanating from the office of the Chief Martial Law Administrator (CMLA) in the form of martial law orders, ordinances, etc. A parallel system of martial law courts and tribunals was set up which impinged on the jurisdiction of the higher courts. At the same time the Islamization of the judiciary took the form of *Shariah* benches in each provincial High Court, followed by the setting up of a Federal Shariat Court (FSC) in 1980. It was entrusted with the task of ensuring that laws were not repugnant to the Quran and *Sunnah*, despite existing provisions to this end in the 1973 Constitution. Appeals against the judgements of the FSC were to be heard by the Shariat Appellate bench of the Supreme Court.

The tenure of FSC judges is three years, which can be extended by the President. The judges can be transferred at any time. Out of eight judges, three are supposed to be *ulema* without any formal legal training.

Among the more vexing problems faced by regimes in contemporary Muslim societies seeking legitimacy by emphasizing their commitment to Islam has been the issue of *riba* (usury). The elimination of usury, which according to many Islamist authorities extends to the institution of bank interest, is a key requirement for an Islamic economic system in the view of most Islamist groups in Pakistan. However, not surprisingly, fiscal laws and other related matters were specifically excluded from the jurisdiction of the Federal Shariat Court – albeit for a limited period of ten years under Article 203 B(c) of the Constitution. This period having expired in 1990, the FSC has recently given a ruling holding *riba* (interpreted here as interest) to be un-Islamic – putting the government in a difficult spot.

However, even earlier, peripheral fiscal and economic measures were taken, such as the cosmetic interest-free banking. Then there was the well-publicized compulsory levy of *zakat* (the Islamic poor rate) in 1980. *Zakat*, a tax on capital, not income, was levied at a rate of 2.5% per annum on savings accounts, government securities, life insurance policies etc. and local committees were set up to distribute proceeds to the poor. However, compared to the overall revenues, this amount was minimal. For instance, in 1984, the *zakat* levy yielded about Rs 1 billion (USD 75 million) compared to the state's other revenues of about Rs 60 billion (USD 4.5 billion): some 2% of the total (Faruki, 1987, p. 63). The strong Shia minority, on the other hand, launched a sharp protest against this mode of instituting *zakat*, contending that according to their *fiqh* (sect) such charity must be collected and distributed in the usual manner and not through the state. The government soon relented, and exempted Shias from paying *zakat* to the state. Nevertheless the funds generated by *zakat* and its distribution have certainly helped to gain support for Zia and his constituency. A substantial amount was distributed through the *madrassas* largely belonging to Sunni sects of Deoband, Ahl-e-Hadith and Barelvi. The number of *madrassas* at the time of partition was 137. By 1971 it had grown to an estimated 893 with a total of 3,186 teachers and 32,384 students (Rahman, 1982, p. 42). Given the funds put at the disposal of religious parties and organizations during the Zia years, the number of students and teachers at such *madrassas* must have increased many times over, supplemented by a number of new institutions.

As part of mainstream educational reforms, Zia called for revising the curriculum and textbooks at various levels starting from the primary, to make pupils more aware of the Islamic mainsprings of Pakistan's creation, its ideology and the history of Islam. In addition, an Islamic University was established at Islamabad with the objective of fusing 'Islamic' and 'secular' education. It was to answer the criticism that the learned in *Din-e-Islam* (Islamic religion) almost invariably lacked modern education and training and that they were hence disadvantaged in prescribing how Muslims should order their lives in this day and age. Its graduates were to staff the *Qazi* courts that would eventually form a parallel 'Islamic' – and supposedly more equitable – system of justice. Significantly, the university was funded largely by Saudi Arabia, with that country retaining a say regarding the choice of the faculty. Also official recognition was given to degrees from traditional theological schools and colleges (*deeni madarris*).

In line with General Zia's wishes, the University Grants Commission issued a directive for the benefit of prospective textbook authors spelling out the objectives of the new curriculum: to demonstrate that the basis of Pakistan is not to be found in racial, linguistic or geographical factors, but rather, in the shared experience of a common religion; to get students to know and to appreciate the ideology of Pakistan, and to popularize it with slogans; to guide students towards the ultimate goal of Pakistan – the creation of a completely Islamized state (Hoodbhoy & Nayyar, 1985, p. 165). On the other hand, some scientists were quick to jump on the bandwagon, making astonishing pronouncements about such things as the nature of *jinns* (genie) and the 'Angle of God'.[4]

In all of this, the two most immediately affected groups in society were women and minorities. Many women were sentenced under the Hadood Ordinances for the crime of *zina*, which made no distinction between voluntary sexual intercourse and rape. More generally their position was undermined through laws such as the *Qanoon-e-Shahadat* (Law of Evidence) promulgated in 1984, which equated the testimony of one man to that of two women. Court rulings helped in reducing the impact of such laws, however. A distinction was made for instance between rape and adultery in the course of a judgement by the FSC. Similarly protest demonstrations led by urban women's groups such as the Women's Action Forum led to modifications. A proposed law submitted by the Council of Islamic Ideology, an advisory body, limited discrimination between men and women to financial matters. Then in 1981 the FSC ruled that there was nothing Islamic about *rajam*, the punishment of death by stoning, in opposition to the government's position. As a result, there came a Presidential Order empowering the FSC to review its own judgements. The court was rapidly reconstituted, with only one judge from the original bench being retained, and the FSC promptly reversed its earlier decision. The FSC has upheld the position that a woman (or for that matter any number of women) is not competent to bear witness in a *zina* case where *hadd* (maximum punishment) is applicable. In such cases, including rape, only the testimony of four male Muslim eye-witnesses 'of good repute' would be considered adequate. Fortunately the law does allow the court to consider the woman's testimony in cases where lesser punishment (*tazeer*) is involved.

Among the minorities the most affected was the Ahmadiya community. Under Bhutto, they had been officially given minority status, but were more actively persecuted under Zia – who by ordinance made it a crime, for instance, for the Ahmadiyas to inscribe Quranic verse (*kalima*) on their places of worship, which were no longer to be referred to as mosques. They also became victims of the vigilante mentality that took hold under Zia.

More generally, the separate electorates introduced by Zia in 1985 alienated the Christians, Hindus, Parsis, Ahmadis and other minorities from the political mainstream. On the other hand during the decade 1978–88 probably more houses of worship of non-Muslims – and among Muslims, *Shia Imam Baras* – were attacked and destroyed than ever before in Pakistan's history.

In the broader context of reorienting the rules and laws governing society perhaps the key constitutional amendment made by President Zia-ul-Haq was the insertion of Article 2A through Presidential Order No. 14 of 1985, making the Objectives Resolution, which had formed part of the Preamble of the Constitution, a substantive part of it.

Leaving aside the Jamaat-e-Islami, which is more modern (as well as fundamentalist) and politicized in its approach, other groups should not be discounted. The Jamiyat-ul-Ulema-e-Islam (JUI) and the Jamiyat-ul-Ulema-e-Pakistan (JUP), the former representing the Deoband and the latter the Barelvi school of *ulema*, have served as politico-religious parties of the more traditionalist variety. Their influence has been exerted through the institutions of mosques and *madrassas*. Others, such as the Ahl-e-Hadith, have become politicized more recently. The latter may be distinguished by its disregard for the classical schools of jurisprudence while it emphasizes a return to the fundamentals of the Quran and *Sunnah*. Ironically, the Zia regime with its emphasis on Islamization and notions of promoting a broader unity, perhaps a merging together of Islamist parties, saw the emergence of new parties that emphasized sectarian differences, such as the Tehrik-i-Nifaz-e-Fiqh-i-Jafaria (TNFJ). It was formed as a result of the widespread perception among the Shia community that Pakistan was becoming an entirely Sunni, even *Hanafi* (a sub-sect among Sunnis), state and that this called for an organized response. This was paralleled in the emergence of the Anjuman Sipah-e-Sahaba (ASS) which has enjoyed a close relationship with JUI. The ASS has adopted a far more virulent and extremist stance than the latter, particularly with reference to the Shias.

The Anjuman Sipah-e-Sahaba lost its two successive heads, among other members, through assassination, in the space of two and a half years. The head of the TNFJ, Allama Arif Hussaini, was also assassinated,[5] as was the Consul General of Iran, who was widely perceived by the non-Shia Islamist parties to be actively supportive of the TNFJ and other Shia groups in Pakistan.

The Jamaat-e-Islami had been the first political party in Pakistan to develop a properly organized and armed cadre of workers, particularly among students. It continued to assert itself, especially in the Punjab. However, during Zia's concluding years, it lost ground in some areas. No longer did it enjoy the same kind of monopoly on campuses, now it was challenged by the Muslim Students Federation (MSF) in some of the larger urban centres and even overtaken, in some cases, by groups like the ASS in smaller ones. In fact, when the JI was being hounded by the Mohajir Qaumi Mahaz (the Mohajir National Movement) in Karachi and complained about scores of their workers murdered by the MQM, it was said that among those responsible were people trained by the JI who had switched sides on the basis of ethnic affiliation with the MQM.

However, JI remains a force to be reckoned with in the Punjab, and to some extent in the North West Frontier Province, partly due to the Afghan nexus. The external arms and money linkages developed by Islamist forces during the Zia years are a factor in this context – not that such assistance was unheard of previously, however. The JI's links with Saudi Arabia had been an open secret for many years, although the Jamaat's new leadership with its more radical pretentions has taken care to play these down. The Iran–Iraq War and the subsequent rifts within the Muslim world also made such patrons more amenable to doling out funds to their favourite Islamist party. The flow of funds to such parties from Saudi Arabia, Iran, Iraq and possibly Libya increased considerably during the 1970s (1973 oil boom and 1979 Iranian Revolution). The JI's position was further strengthened by its culti- vating an extremely close relationship with the Hizb-e-Islami of Gulbadin Hikmatyar, the Peshawar-based extremist Afghan group. The Hizb, through the Inter-Services Intelligence of Pakistan, was the greatest single beneficiary of the massive covert arms supply to the Afghan resistance over a number of years. The relationship is seen as a potential asset for the JI.

The Jamiyat-Ahl-Hadith was also a major beneficiary of external assis- tance from Arab countries, including Iraq. The assassination of its leader, Ehsan Elahi Zaheer, was alleged to be the work of a rival religious group.

In order to demonstrate broad social support for his Islamization enter- prise, Zia held a referendum on the issue, seeking at the same time a mandate for the continuation of his policies and linking 'Islamization' to his own continuation in office. That most independent observers pegged the turnout at a dismal 10% despite strenuous efforts by the administration did not deter Zia from claiming that the referendum had indicated overwhelm- ing popular support. The non-party polls which he held subsequently did elicit impressive popular response, despite a boycott by the major political parties, except for Jamaat-e-Islami. Whatever the reasons, these polls had very little to do with Islam, as most candidates campaigned around local issues. A particularly unfortunate outcome of these polls was that, in the absence of political parties, ethnic, sectarian and *biradri* (kinship) divisions became reinforced in the course of the campaign. Along with Zia's other efforts at depoliticization this was to have serious long-term implications, particularly for Sindh.

4.5   Building on the Zia Legacy

The drive for Islamization slowed down somewhat with the lifting of martial law and induction of Prime Minister Junejo into the system in 1985. A Shariah Bill was introduced in Parliament in 1985 by Senator Samiul-Haq, followed by the Constitution (Ninth Amendment) Bill, along similar lines. Neither became an Act of Parliament, however. When the assemblies were dissolved by President Zia in May 1988, momentum was sustained through the passage of a Shariat Ordinance.

It was eventually in 1989, under the Benazir Bhutto government, that a largely hostile Senate passed a modified version of Senator Samiul-Haq's Shariah Bill, to the obvious delight of President Ghulam Ishaque, who congratulated the Senate on their accomplishment. In its approximately 20 months in office, the PPP did show an inclination to roll back the Ziaist Islamization but it was unable to make much headway. There was of course the hostile Senate. But the strength of the Ziaist legacy was best illustrated by the induction into office of President Ghulam Ishaque, an experienced bureaucrat and one of Zia-ul-Haq's key advisers. He quickly made clear his own orientation and preferences, including one for the continuation of Ziaist Islamization. Another telling feature was the continuation in office (as Chief Minister of the Punjab, the country's most populous province) of Nawaz Sharif – indebted to the generals for his elevation and in the vanguard of opposition to the PPP. However, this Shariah Bill, passed by the Senate, later lapsed when the assemblies were once again dissolved, this time by President Ishaque, in 1990.

The issue of the Shariah Bill was taken up with gusto again when Nawaz Sharif became Prime Minister in 1990 in the wake of elections whose fairness was questionable. That the Bill became a live issue again had to do with the JI being a key coalition partner in the Nawaz Sharif-led coalition Islami Jamhoori Ittehad (IJI), Sharif's own desire to tap the Zia constituency, and the fact of President Ghulam Ishaque's identification with the Bill and the Islamization project generally. It should be kept in mind here that the 1973 Constitution as amended by Zia confers, in effect, the powers of the chief executive on the President. Such powers have already been used twice to dismiss elected Assemblies in an arbitrary manner, once by Zia in 1988 and once by Ishaque in 1990.[6]

Prime Minister Nawaz Sharif's government passed its Shariah Act in 1991, six years after the first such initiative in the Parliament. Technically, the Act cannot take effect without an 'enabling' constitutional amendment which has yet to be passed by Parliament. The main import of the Act lies in the direction which it marks out for the legislature and the government with regard to Islamization, bringing to bear greater pressure to this end.

On most issues the Act remains vague. For instance, it commits the state to take 'steps' for 'Islamization of the economy' ('Enforcement', 1991, clause 8), take 'effective steps' to Islamize education ('Enforcement', clause 7), etc. Commissions have been appointed with respect to these key areas to make necessary recommendations towards these ends. That deadline is indicative of a desire on the part of the government to keep the initiative in its own hands. Meanwhile clauses 18 and 19 of the Act specifically exempt Pakistan's international financial dealings from the constraints of the Shariah Bill until an 'alternative economic system' is evolved. Fortunately, the Act also protects the rights of women as guaranteed by the Constitution ('Enforcement', clause 20), which would make it difficult for the Family Laws Ordinance to be challenged under this act.

There is considerable uncertainty as to what may or may not be successfully challenged if the enabling constitutional amendment does make the Shariah, in an operative sense, 'the supreme Law of Pakistan' ('Enforcement', clause 8). This could make it possible for judges (appointed and transferred by the

President) to place their own interpretation (while following 'recognized principles of interpretation . . . and the expositions and opinions of recognized jurists of Islam') above constitutional provisions. Technically, then the Constitution would not remain the supreme law of the land.

Again with regard to the issue of interpretation, the Act reaffirms the centrality of religious sects, a concept first introduced in the Constitution of Pakistan in 1980 in the form of an explanation to Article 227, clause 1, which reads 'All existing laws shall be brought in conformity with the injunctions of Islam as laid down in the Holy Quran and Sunnah . . . and no law shall be enacted which is repugnant to such injunctions.' In the explanation it has been spelt out that 'in the application of the clause to the personal law of any Muslim sect the expression Quran and Sunnah shall mean the Quran and Sunnah as interpreted by that sect'. Clearly this reinforces existing divisions and makes consensus jurisprudence a more remote possibility.

Amidst all this, however, there is little doubt regarding the extent to which the country is dependent on external sources of military and economic assistance. This was neatly illustrated in one brief sentence during the course of a rambling speech in Urdu by Prime Minister Nawaz Sharif. Announcing the passage of the Shariah Bill in the National Assembly, on the national media, he declared (in English) 'I am not a fundamentalist.'

The dilemma for the government then is how to retain and maximize the power in its own hands, placate and use religious groups in this exercise and at the same time not to give away too much or appear threatening or 'fundamentalist' to much-needed, non-Muslim (external) sources of support. Meanwhile, other laws based on traditionalist views of Islam are being put on the statute books, dangerously distorting the social balance. There is, for instance, the *Qisas* and *Diyat* Law. The Penal Code was amended to make murder into a compoundable offence under this law. The heirs of a murder victim can forgive the accused who is charged with the offence, either without any consideration or after receiving blood money. In that event, the court has no option but to set the killer free. For a long time, apprehension had been expressed that this provided an opportunity for the rich and the powerful either to frighten the victim's family into granting forgiveness, or to buy their way out.

This happened in a well-publicized case in 1991. A special court for the Suppression of Terrorist Activities in Lahore set free 15 accused who were facing trial for the murder of a Jhang MNA, Maulana Easar Qasimi. They had been forgiven by the victim's families. The murder had sparked widespread rioting in Jhang, already a trouble spot and a centre of sectarian strife. At least two people were killed, property was destroyed and the Maulana's murder itself fanned sectarian strife in Jhang (*Viewpoint*, 1991, p. 21).

The attending circumstances in this case serve to highlight the ability of the rich and the powerful to get away with murder – literally. They also illustrate the grave implications of converting a serious crime against society to one against an individual, particularly in a society as stratified and riven with inequity as that of present-day Pakistan.

4.6 Liberal Secularism and the Politics of Religion

At the same time, the ruling coalition continues to strengthen itself at the expense of the individual and the political opposition in more 'secular' ways through amendments to the 1973 Constitution. For instance, the Eighth Amendment to the 1973 Constitution, pushed through the National Assembly under pressure from Zia in 1985, amended Article 270-A, giving protection to all presidential orders, ordinances, martial law regulations and orders. 'All other laws made between the 5th day of July, 1977 and the date on which this article comes into force . . . shall not be called into question in any court on any ground whatsoever.' Numerous changes in the Constitution invested the President with extraordinary powers that effectively turned a parliamentary system into a presidential one. Similarly, the Twelfth Amendment that was rushed through the National Assembly undermines fundamental rights and provincial autonomy by making the procedures and verdicts of Special Courts more difficult to challenge in ordinary courts via normal judicial processes.

The ruling elites continue to employ a combination of 'Islamic' and 'secular' innovations and 'reforms' to maximize their own power. The purpose is to avoid alienating the well-organized and potentially troublesome Islamist forces on the one hand, and the even more powerful wings of the army, bureaucracy and the bourgeoisie which remain secular, especially in their desire to emulate the West.

More broadly, the misuse of religion and the increasing propensity in various quarters to pursue political ambitions in the guise of religion should not obscure the reality that for the ordinary men and women of Pakistan, Islam does represent a set of cherished values that they will not see violated with impunity, even if at times this becomes an exercise in 'moral double bookkeeping'. On the other hand, they are not easily swayed by religious sloganeering as is evident from the inability of the politico-religious parties to form the government on their own by winning enough votes in a general election. The practice of Islam as a worldly or 'secular' religion makes it more possible for people to ignore narrow theological formulations and support more convincing proposals that address their worldly needs and aspirations without going against the spirit of their religion.

In 1947, for instance, the people chose Jinnah's Pakistan over an almost unanimous verdict by the *ulema* of united India that its creation would serve neither the interests of the Muslims of India nor that of Islam. In 1970 Bhutto was elected in West Pakistan by a landslide victory despite the massive opposition of the religious groups and politico-religious parties. It is in fact mostly due to the failure of the liberal/secularist project in Pakistan that politico-religious forces and the 'Islamist' establishment have been strengthened.

Since most Muslims live with a loose and vaguely defined notion of the *ummah*, events in other Muslim countries, being Islamic, can have a significant impact at the level of popular consciousness. There was, for instance, popular agitation against the government's position in Pakistan during the

1956 Suez crisis. The US embassy in Islamabad was burned in 1979 and there was anti-Rushdie agitation in 1989. Most recently, the savage destruction of Iraq at the hands of the US-led coalition provided politico-religious parties with an ideal opportunity to put themselves at the head of a wave of anti-US and pro-Iraq sentiment in what came to be seen by the public as a Christian crusade against a lone, defiant, Muslim power. The secularists' linkages with the West and acceptance of the latter's hegemony, not only in material domination but moral and intellectual leadership as well, often put them out of step with popular response to developments such as those in Afghanistan and Kashmir, where fundamental rights have been violated and repression unleashed on a major scale. Whether the advantage taken by Islamist parties in such circumstances can be converted into long-term electoral strength remains to be seen. For the present, the external dimension has a strong bearing on 'Islamist' politics in Pakistan.

The conflict in Kashmir gives no sign of waning, and parties such as the JI are now clearly focused on it for their own purposes. Similarly, the Babri Masjid issue in India provides an emotional charge for their agenda. Not least, the rise of Hindu 'fundamentalism', symbolized by the meteoric rise of the Bharatiya Janata Party (BJP), undermines the liberals' argument in Pakistan for keeping religion effectively separate from the affairs of state.

Then there is the intensifying Iranian–Saudi rivalry to bring Pakistan in as part of a broader security arrangement in the aftermath of the Gulf War. As a result, the politico-religious parties in Pakistan, backed by one or the other of the two countries, are likely to be strengthened as the rivalry increases or deteriorates and expands into Afghanistan as well.

In addition, the very 'secular' issue of Pakistan's nuclear programme could have significant, indirect implications for 'Islamist' politics. For instance public perceptions that the programme is being jettisoned under pressure from the West could easily be translated into the religious language of a Christian conspiracy against Islam. Those who argue that the acquisition of maximum power is the only meaningful response in the continuing struggle with the demonstrably implacable enemies of Islamic civilization could be strengthened.

4.7 Another Turn in the Political Path

In October 1993 the PPP-led alliance formed the government in Pakistan under Prime Minister Benazir Bhutto, dealing the politico-religious parties a major setback. The Jamaat-e-Islami, for instance, ended up with a greatly reduced number of seats in Parliament and in the provincial assemblies as well. Partly it seemed that the JI, under its dynamic and ambitious leader Qazi Hussain Ahmad, had over-reached itself in seeking to establish itself as an independent, electorally significant force under the rubric of the Pakistan Islamic Front (PIF). The PIF, the Islami Jamhoori Mahaz and the Mutahida Deeni Mahaz (the latter were alliances made up of politico-religious parties), together managed to secure 1.3 million votes or 6.7% of the nationwide voting figure. In Punjab the PIF vote was 2.4%. It had parted company with its erstwhile ally the Pakistan Muslim League (Nawaz group), also known as the PML(N) which, as it turned out, managed to do very well in the elections and was only narrowly edged out of first place by the PPP.

Nevertheless the politico-religious parties emerged with some clout after the elections. The JI was courted by the mainstream political parties, including the PPP and the PML(N), for its votes during the presidential elections, the senate elections, the election of the senate Chairman and, not least, in the tug-of-war between the PPP and the PML(N)-ANP (Awami National Party) in the Frontier Province over the office of the chief minister. It is indicative that the PIF vote made a difference between a PPP and a PML(N) victory in at least ten National Assembly seats in the Punjab. The JUI(F), which has a close relationship with the Anjuman Sipah-e-Sahaba Pakistan, the most virulent of the lot, also has an understanding with the government. Its Secretary-General, for instance, was appointed the head of the Parliamentary Committee for Foreign Affairs and was one of those sent abroad by the government to plead the case of Kashmir on the eve of the meeting of the Human Rights Commission in Geneva early in 1994, which considered a resolution by Pakistan censuring the gross human rights violations in Kashmir.

The PPP under Benazir Bhutto is generally considered to be more enlightened in its outlook on religion. Ms Bhutto herself has often been attacked by the clergy for saying, for instance, that she was against brutal punishments such as cutting off a thief's hand, etc. However the fact remains that her ability and that of the PPP government is limited in the context of changing an environment of religious intolerance and sectarianism which owes a lot to the Zia years – as well as the propensity of the Sharif-led PML(N) to conform or complement rather than seek to counter it. At one level, of course, the PPP may feel hemmed in as a result of the stark arithmetic of electoral politics in the country. Despite their reduced strength, the politico-religious parties are still in a position to help tilt the balance between the two major parties on occasion. But more importantly, in a social milieu comprising well-organized, armed and often well-financed politico-religious parties and groups, insistent on carrying forward many of the changes brought about by Zia for essentially self-serving political ends, the room for manoeuvre is limited. Especially when the opposition lends them support, tacitly or otherwise, in the context of their own efforts against the PPP government. Consequently, laws that are discriminatory and unjust, as well as being contrary to the spirit of Islam, such as the Hadood Ordinance, blasphemy laws etc., remain on the statute books.

A law that has become the subject of controversy recently has been Section 295-C: 'Use of derogatory remarks, etc., in respect of the Holy Prophet: Whoever by words either spoken or written, or by visible representation, or by imputation, innuendo or insinuation, directly or indirectly, defiles the sacred name of the Holy Prophet Muhammad (peace be upon him) shall be punished with death, or imprisonment for life, and shall also be liable to fine.' This section was added in 1986 (following Sec. 295-B in 1982). A Shariat court ruling under the Nawaz Sharif government did away with the option of life imprisonment and made capital punishment mandatory for anyone convicted under Sec. 295-C. The minorities, particularly the Christians, are now making a greater effort to protect themselves from victimization under such laws. While threats of resigning from the legislature and demonstrations are likely to have an impact, some of the steps may be misdirected. In what appeared to be a bid to acquire a measure of 'deterrence' for the Christian community, a Pakistani bishop went to court and had it acknowledged that all Prophets, not just Mohammad (peace be upon him), are covered by the blasphemy laws. However, while societal pressures are unlikely to allow the law to become an instrument of leverage in the hands of minorities to work against members of the majority community, the latter may now feel more free to take up the issue of how, for instance, Jesus, who is also a prophet for the Muslims, is portrayed by the Christians. In any case, the largest number of those affected have been Ahmadis who have been before the law for misrepresenting themselves as Muslims.

Muslims, though affected in much smaller numbers, have not been immune to the reach of the blasphemy laws. Several people, including a highly respected and internationally known social worker, Akhtar Hameed Khan, have been implicated in such cases. Judges at various levels are reluctant to let an accused go free in such cases, perhaps feeling apprehensive that they may be regarded as being soft on the 'enemies of Islam'. At the level of the lower courts, the threat is more palpable, as hearings are usually attended by charged mobs wanting 'justice' done without further delay. The law has been frequently misused by unscrupulous elements to settle scores or resolve disputes by getting the other side involved in such a case and then letting the courts – or an obliging mob – do the rest. But it is important to keep in mind that only rarely has a court actually convicted people and sentenced them to death. Certainly, no one has been put to death legally as a result of such a process. But the existence of such laws tends to create an environment of hysteria in which, once such an allegation has been made, few are likely to examine the merits of the case or take a stand against the charge.

In May 1994, the Bhutto government declared that the cabinet had given its approval to two amendments in blasphemy law 295-C. In the first, the police could only register a case under this law after a competent court had ascertained that there was enough substance to warrant such registration. Secondly, anyone making false allegations would be liable to the severe punishment of a ten-year prison term. This step was meant to discourage the casual registration of such cases as a way of settling scores or getting someone out of the way.

This change is clearly a step in the right direction. But two points need to be noted here. It has come after massive protest by the Christian community in the wake of the daylight murder of an accused in a blasphemy case involving Manooor Masih, who was on bail and had just left the court after a hearing. At about the same time Sajjad Farooq, a *Hafiz-e-Quran* (one who knows the Holy Quran by heart), was stoned to death by a mob within hours of being accused of burning a copy of the Holy Quran. The proposal for amending the law, therefore, came at a time when the government could perhaps be more confident that such a move could not be branded as being anti-religious. However, the government still did not feel that the law could be set aside in its entirety.

The government will not find it a simple matter to alter the extent to which the institutional setting now reflects such narrow-minded religiosity. The Federal Shariat Court set up by Zia, with many of the judges inducted or promoted during his time or that of Mian Nawaz Sharif, and the manner in which other institutions were 'refashioned', have had an enduring impact. A further complication is that the PPPs junior coalition partners, the Pakistan Muslim League-Junejo group – the PML(J) – probably have more in common with their former colleagues of the PML(N), regarding such matters, and are hardly likely to be enthusiastically supportive of changes aimed at rolling back this aspect of the legacy of the Zia years.

4.8   Conclusions

The 'fundamentalist' position in Pakistan, though well-articulated by the Jamaat-e-Islami, has lacked the emotional charge and credibility that could turn it into a popular movement. The JI remains an important factor in Pakistan's politics, not because of the numbers it commands, but by virtue of its ability to strengthen or undermine governments through its hold among the *petit bourgeoisie*, its presence in the civil and military bureaucracy and its manipulation of certain trade unions, campuses etc.

Similarly, in General Zia-ul-Haq's 'administrative' and downward-directed Islam, 'fundamentalism', with its emphasis on debatably Islamic punishments, was popularly seen as little more than an exercise in the pursuit of power and continuation in office through the coercive use of the state apparatus and ideology. It was for this reason that the referendum held by Zia – linking the people's preference for some kind of an 'Islamic' system to his own continuation in office for another five years – was seen as a farce by the overwhelming majority of the people. Despite massive efforts by the administration, less then 10% of the people are estimated to have voted in what was supposed to be a dramatic reaffirmation by the masses of 'Zia's Islamization'. It must be added, though, that through his various measures, Zia certainly strengthened the religious establishment in Pakistan, even as he aggravated the divisions among its various components. Under his rule the division between the two main sects, Shia and Sunni, became formalized along party lines. Several existing Sunni politico-religious parties were put into sharp sectarian relief by the emergence of the Shia TNFJ. Another Sunni politico-religious party to emerge at the time was the ASS.

It is a kind of comment on the nature and extent of the sectarian division that both parties have lost their top leadership to murder and assassination

and have accused the opposite sect. While both can be seen as 'fundamentalist', perhaps the more appropriate description would be sectarian extremism.
Similarly other politico-religious groups, currently less powerful and effective,
are a part of the Zia legacy. These include Pakistan Awami Tehrik (PAT) of
Maulana Tahirul Qadiri and the Islami Tehrik of Maulana Israr. Both have
made an impact among the bourgeoisie and have, like the JI, a degree of following among middle-level civil and military bureaucrats. While all such
politico-religious groups are unable, singly or collectively, to form a government by winning an election (without allying themselves to their more
'secular' counterparts), or to take on the state apparatus in a direct confrontation, the last decade has definitely seen a substantial increase in their
impact on civil society, as well as their bargaining position vis-a-vis the state.

Meanwhile the failure of the secular (rather than liberal) establishment to
succeed is made apparent by Pakistan's acute economic crisis. This has been
aggravated by rampant corruption, the rise of a powerful arms and drug
mafia (a consequence of Pakistan's covert role in the Afghanistan war), and,
more generally, the inability of Pakistan's ruling elite to formulate a grand
design or a strategy of development which can satisfy the great majority.

This failure is most dramatically illustrated in the parallel strains of extremism and divisiveness that have manifested themselves along the (secular?)
lines of ethnic reassertion in the country's most industrialized province,
Sindh, and in its most modern urban centre, Karachi. While ethnic consciousness in Sindh is nothing new, it took Zia's martial law to widen the gulf
between the different ethnic groups to such unprecedented proportions. Thus
emerged the MQM, the Pakhtun Punjabi Ittehad (PPI) and the Jeay Sindh
(JS), for the first time coming into its own as a significant political entity.

In such a state of crisis, confusion and deeply felt uncertainty, it is indeed
possible for a fundamentalist creed to gain acceptance at a national level.
However, sectarian divisions and the lack of credible leadership make such a
prospect unlikely. For the present it appears that essentially secular institutions
such as the army, along with powerful sections of the feudal and bourgeois
elite, will still call the tune, and that 'Islamization' in some form will continue
as well. Also likely to continue is an ongoing process of negotiation and compromise between the politico-religious forces and those representing the state
and the modern sector – probably to the detriment of the spiritual ethos, as
well as the aspirations of the people for political freedom and material welfare.

Clearly none of this is heading in the direction of an authentic and indigenous liberalism which can become the vehicle for addressing the central
issues of democracy and development in Pakistan. Only the elaboration of
such a framework will make it possible for the universal values associated
with liberalism to be appropriated and internalized, rather than superficially accepted or adopted by an elite. The contribution of philosophy and
Sufism must again form part of the Islamic discourse, widening the narrow
limits of traditionalist and fundamentalist doctrines.

At one level the exclusion of rationalists, like the Mutazila, from the mainstream of Muslim thought has to be contested. At another, one must build

upon the rich and pervasive tradition of humanist Sufi poetry in the Punjab and Sindh, for instance. The inspiration for a system that upholds democracy and rationalism has to be derived not only by recourse to religion or faith, but also to that which the people perceive to be a part of their heritage and history: neither of which, in the case of Pakistan, is exclusively 'Islamic'. But, on the other hand, it would be difficult to exclude Islam entirely from the cultural and historical matrix.

Not least, what modernist efforts at interpretation have lacked is an element of inspiration. Apart from the absence of an interpretative methodology, as emphasized by Fazal-ur-Rahman, such initiatives have not demonstrated a degree of autonomy that would set them apart from the orientalist perspective of the West. The separation of the notions of Westernization and modernization – fused in the colonial history of Muslim societies such as Pakistan – will have to accompany the elaboration of an indigenous liberalism.

Notes

1. For an interesting discussion of Khomeini as a populist phenomenon, see Evrand Abrahamian, (1991).

2. A major exception was Maulana Shabbir Ahmed Usmani, who led a breakaway faction of the Jamiyat-ul-Ulema-e-Hind in support of Pakistan, subsequently Jamiyat-ul-Ulema-e-Islam.

3. Notwithstanding the way it was presented subsequently and the construction put upon it by those who wanted to read it as a liberal document, it would be useful to reflect on the remarks of the Prime Minister, Liaquat Ali Khan (quoted in W.C. Smith, 1957 p. 215). While introducing the resolution, he said 'The State is not to play the part of a neutral observer wherein the Muslims may be merely free to profess and practise their religion, because such an attitude on the part of the state would be the very negation of the ideals which prompted the demand of Pakistan, and it is these ideals which should be the cornerstone of the state which we want to build. The state will create such conditions as are conducive to the building up of a truly Islamic society, which means that the state will have to play a positive part in this effort.'

4. See Pervaiz Hoodbhoy, (1991, p. 174). He quotes, to illustrate all that was marketed by the Zia regime in the name of Islam, from the presentation of a German delegate (which interestingly went unchallenged) to the Islamic Science Conference held in Islamabad in 1983 where the gentleman claimed to have calculated the 'Angle of God' mathematically as pi/N with the value of pi being 3.1415927 and N left undefined.

5. Recently the former governor of the Frontier Province, Lt General (Retired) Fazle Haq, alleged by many of Hussaini's followers to have had a hand in his death, was also gunned down near his home in Peshawar.

6. Article 58 of the 1973 Constitution read, 'The President shall dissolve the National Assembly if so advised by the Prime Minister . . .'. This was made Article 58(1) and a Clause 2 added by Presidential Order No.14 of 1985 which read: 'The President may also dissolve the National Assembly *at his discretion* [emphasis added] where, in his opinion, an appeal to the electorate is necessary.'

References

Abbot, Freeland, 1968. *Islam and Pakistan*. Ithaca, NY: Cornell University Press.

Abrahamian, Evrand, 1991. 'Khomeini: Fundamentalist or Populist?', *New Left Review*, no. 186, March/April, pp. 102–119.

Ahmad, Aziz, 1967. *Islamic Modernism in India and Pakistan*. London: Oxford University Press.

Alavi, Hamza, 1986. *Pakistan and Islam: Ethnicity and Ideology*. Paper presented at the 9th European Conference on Modern South Asian Studies, University of Heidelberg.

Baxter, Craig, Yogendra K. Malik, Charles H. Kennedy & Robert C. Oberst, 1988. *Government and Politics in South Asia*. Boulder, CO: Westview Press

Faruki, Kemal A., 1987. 'Islamic Government and Society', pp. 53–78 in John Esposito, ed., *Islam in Asia*. New York: Oxford University Press.

Government of Pakistan, 1954. *Report of the Enquiry Constituted Under Act 11 of 1954 to Enquire into the Punjab Disturbances of 1953*. Lahore: GOP.

Government of Pakistan, *The 1973 Constitution*. Karachi: Manager of Publications.

Hoodbhoy, Pervaiz, 1991. *Muslims and Science*. Lahore: Vanguard Books.

Hoodbhoy, Pervaiz & Abdul Hameed Nayyar, 1985. 'Rewriting the History of Pakistan', pp. 164–177 in Asghar Khan, ed., *Islam, Politics and the State: the Pakistan Experience*. London: Zed Press.

Iqbal, Muhammad, 1960. *The Reconstruction of Religious Thought in Islam*. Lahore: Sheikh Mohammad Ashraf.

Khan, Syed Ahmad, 1873. *The Causes of the Indian Revolt*, 8th edn. Benares: Medical Hall Press.

Maududi, Abul Ala, 1981. *Islamic State*. Lahore: Islamic Publications.

Mortimer, Edward, 1982. *Faith and Power*. London: Faber & Faber.

Noman, Omar, 1990. *Pakistan Political and Economic History since 1947*. London: Kegan Paul International.

Rahman, Fazal-ur-, 1979. 'Islam: Challenges and Opportunities', pp. 315–330 in Alford T. Welch & Pierre Cachia, eds, *Islam: Past Influence and Present Challenge*. New York: State University of New York Press.

Rahman, Fazal-ur-, 1982. *Islam and Modernity*. Chicago, IL: University of Chicago Press.

Smith, W.C., 1957. *Islam in Modern History*. London: Oxford University Press.

Smith, W.C., 1969. *Modern Islam in India*. Lahore: Ashraf Press.

5

The Politics of Violence in the Indian State and Society

SUMANTA BANERJEE

5.1 Introduction

Among post-colonial independent states in South Asia, India seems to have emerged as the classic example of a state trapped by a built-in contradiction. The contradiction is between its self-proclaimed ideological basis and the objective reality of its actions which consistently violate its professed ideology.

This contradiction is built-in because its roots are historical. Official spokespersons – as well as official historians – of the Indian state would have us believe that Independence was brought about through a non-violent mid-wife called 'Gandhiism', which is the state's proclaimed ideology. Yet history reveals that the most violent birth-throes accompanied the foundation of the independent Indian state. The Hindu–Muslim communal riots that preceded the partition of the subcontinent and led to the birth of 'India, that is Bharat', are a historical reality which not only knocks the bottom out of the Indian state's official and ideological claim to its 'non-violent' parentage, but also challenges the doctrine of non-violence as propounded by Gandhi as a har-binger of change. In fact, towards the end of his life, Gandhi sought to distance himself from the violent political midwifery which had delivered the twins – one called India, the other Pakistan – from a tortured womb. His iso-lation from the mainstream, tinged by the realization of the failure of his life-long dedication to the doctrine of non-violence at the end of his political career, has been amply documented (Bose, 1962). But that is another story!

In talking about militarization in India we need to remember that it cannot be considered in isolation from the fact that the Indian state was born of vio-lence, and that violence still continues to dominate its society. This attempt to link the violence that accompanied the birth of the Indian state with the vio-lence that marks present Indian society might be misconstrued by some as an expression of some irrational belief that the state is doomed to violence because it was born under an evil star. But a dispassionate assessment of his-tory will reveal that the violent outbursts that are tearing apart the fabric of Indian society today are expressions of the same contradictions – religious, class, ethnic, caste and otherwise – that often exploded into violence during the colonial period, and whose extreme manifestations were the pre-partition

communal riots.[1] The independent Indian state inherited these contradictions from the pre-1947 era, and has failed to resolve them. Despite the
Indian state's declared profession of adherence to *ahinsa* or non-violence
(which it invokes from historical precedents like Buddha and Ashoka – down
to Gandhi – by formulating official symbols and rhetoric that represent their
doctrines), in practice the Indian state has repeatedly resorted to the language of violence in trying to resolve these contradictions that plague Indian
society.

The arguments presented here can be formulated along the following lines:

1   The very ideological foundation claimed by the Indian state (Gandhian
    non-violence) was flawed from its birth.
2   Since then, there has been continuity in what we can describe as a strong
    streak of violence, despite its ebb and tide and notwithstanding the dif-
    fering causes for such violence over the past four decades. These causes
    have stemmed from pre-Independence contradictions – religious, regional,
    ethnic, linguistic, casteist and class-based – which became accentuated in
    the post-Independence period.
3   It is important to emphasize this *continuity of violence* in Indian politics in
    the context of the *continuity of a political system* and a Constitution ded-
    icated to eradicating the contradictions mentioned above. The violence
    that marks Indian society today cannot be traced to any structural
    changes in the political system (as may be evident in Sri Lanka, Pakistan
    and Bangladesh, all of which experienced a series of tumultuous changes
    in the post-colonial period), which in India remained stable all these years.
4   The stability of the Indian political system has been further reinforced by
    a well-organized armed apparatus (consisting of the police, paramilitary
    forces and the army) which is used by the state to suppress internal dis-
    turbances as well as in its efforts to establish hegemony in South Asia (e.g.
    India's intervention in the Bangladesh war of independence in 1971, and
    again in Sri Lanka in 1987).
5   While the accentuation of violence within India can be traced to the
    state's increasing propensity towards monopolization of power at the
    Centre (which agitates the internal contradictions to danger-point), the
    acceleration in militarization at the external level is prompted by the
    state's attempt to establish hegemonistic power in the South Asian sub-
    continent.

5.2   Militarization – External Level

The Indian state acquired a military character almost from its infancy. It
started at two levels – external and internal. At the external level, the new-
born state sent its troops to Kashmir in October 1947 to fight first the
Pakistan-aided tribals, and later the regular forces of the Pakistan army, in
order to protect a territory which had been acceded hurriedly to the Indian

Union by the then Maharaja of Kashmir. The question of accession still remains a bone of contention between India and Pakistan, with the people of Kashmir yet to have a chance of expressing their independent option as to whether they want to remain a part of the Indian Union or of Pakistan (which is in control of nearly 32,000 square miles of Kashmir's original 36,000 square miles), or have independent status. Despite the ceasefire agreement between Pakistan and India signed on 1 January 1949, the Kashmir imbroglio dominated the two Indo-Pak wars fought so far. Furthermore, it continues to contribute to the militarization of the Indian state which deploys contingents of its numerous paramilitary forces – the CRPF (Central Reserve Police Force), BSF (Border Security Force) – as well as regular armed troops in the Kashmir valley to quell a secessionist movement there.

A series of misadventures and miscalculations made by the central ruling powers in their design to keep rigid control over the course of political developments in Kashmir led to the erosion of what little autonomy had been granted earlier to the Kashmiris and to the gradual estrangement of the people there from the Centre.[2] Meanwhile, Pakistan, taking advantage of the disgruntled among the Kashmiri masses, has been making inroads into the secessionist movement there by offering it military aid. This has led India to become embroiled in a perpetual armed confrontation with Pakistan. The Indian state is thus paying the price for having ignored in the past the need for a democratic solution to the problem of the status of the Kashmiri people.

Three major wars followed the 1947 Kashmir adventure – the 1962 war with China, and the 1965 and 1971 wars with Pakistan, in the course of which the militarization of the Indian state grew by leaps and bounds. From ten divisions in 1962, the strength of its army went up to 21 divisions in 1965, and further to 25 in 1971. By 1988, there were 37 divisions, and the figure is expected to reach 50 by the year 2010 (Rikhye, 1990). Similarly, annual defence expenditure soared up from Rs 3,125 million in 1961–62 to Rs 1,570,500 million in 1989–90 – a fifty-fold increase in the course of three decades. Apart from the three wars, the extra-territorial military adventure in Sri Lanka has cost India dearly, in terms of both army manpower and military prestige. Three years of the so-called peace-keeping operation there left the Sri Lankan Tamils (whom the Indian Peace Keeping Force was supposed to protect) thoroughly hostile to the Indian troops, which had to retreat without achieving the original goal of bringing 'peace' to the war-torn island. Instead, they carried back home the bodies of their dead comrades, maimed by LTTE mines.

The entire Sri Lankan episode epitomizes the Indian state's pathetic search for recognition as a mini-superpower in the subcontinent. It is in this ambition that the roots of militarization lie. In 1962, Nehru nursed the naive notion of swiftly defeating the Chinese, which set India on a war which need not have taken place and which imposed on it a humiliating defeat from which it is still to recover. Ever since then, the Indian state has been trying to beef up its muscles – which means an increasing pace of militarization. The 1965 war with Pakistan started with the fanfare that Sialkot and Lahore were within easy reach – a piece of propaganda that helped to rally the Indian middle-class public who were still smarting under the humiliation of defeat three years before. But to no avail. The Tashkhent Agreement ended in stalemate: India gained nothing in concrete terms, and was left with yet another tally of dead and maimed servicemen.

In 1971, Indira Gandhi was hailed on the walls of Calcutta graffiti as the 'rising sun of Asia's freedom'. This was after she had intervened in the civil war in what was then East Pakistan, and supposedly cut Pakistan down to size by dismembering the state and creating Bangladesh. This inflated the ego of Indian militarism. But again, what was the final outcome? Within four years, Bangladesh was back to square one, with the army taking over, leaving the Bangladeshis in the same plight to which they had become used during the Pakistan regime – with the only difference that now their rulers were Bangladeshi army generals, some of whom (like the ousted General Mohammad Ershad) had received their training in military academies in India!

One of the fall-outs of the 1971 Indian military intervention in East Pakistan appears to be an increasing belligerency on the part of Pakistan. An Islamabad suffering from the humiliating defeat of 1971 must have its revenge. The only way to carry this out is by paying back India in the same coin – in other words, fishing in India's troubled waters. Just as India sheltered, trained and armed the 'freedom fighters' of East Pakistan, Pakistan now feels justified in doing the same with the Khalistani and Kashmiri insurgents who have taken shelter on its soil.

But the problem of violence in Punjab and Kashmir cannot be simply shirked off as Pakistani machination. Even if tomorrow New Delhi should arrive at a rapprochement with Islamabad and succeed in cutting off the insurgents in those two states from the source of their military aid in Pakistan, the armed secessionist trends there are likely to continue, albeit less vehemently for some time perhaps. We have seen in the past how despite China's cutting off aid to the Naga insurgents (following a thaw in Beijing–New Delhi relations), the underground armed Naga movement, after a brief lull, re-emerged as a force to reckon with in Manipur and elsewhere in the north-east. The source of the basic problem therefore is not some external enemy – be it China or Pakistan – but internal. Those troubled waters in which the external enemy likes to fish are India's own creation.

5.3   Internal Sources of Violence

This brings us to the internal sources of violence and militarization in India. They can be traced to the highly centralized model of development adopted by the post-Independence Indian state, which instead of eliminating semi-feudal production relations in large parts of the country opted for investments in selected sectors and regions. As a result, while inequitable distribution of income has sharpened class fragmentation and contradiction, uneven development has led to imbalances in regional growth (Prasad, 1989). The perceived sense of injustice among the people is becoming fissured along the lines of caste, ethnic loyalties, language, religious beliefs – differences and divisions which the Indian polity has failed to overcome, despite its commitment to a 'socialist secular democratic republic'.

The Telengana Movement of 1948 was one of the earliest indications of the Indian state's attitude towards the underprivileged and of its determination to suppress violently any attempt by the rural poor to end outmoded social and economic relations. In the course of a guerrilla struggle against the ruling Nizam of the then princely state of Hyderabad in south India, poor peasants under the leadership of the undivided CPI (Communist Party of India) from 1946 to 1948 liberated wide areas of Telengana. They then distributed land among the landless, putting an end to the domination of the feudal landed gentry. In 1948, the Indian government sent troops to Telengana to take over the state of Hyderabad, in the process unleashing a reign of terror on the peasants and striking at the land reforms initiated by the CPI. To quote one of the participants (Reddy, 1973, p. 60):

> People were made victims of severe violence and repression . . . They were beaten with lathis and bayonets and tortured to the extreme – like peeling the skin in the design of the hammer and sickle . . . Arrested comrades were tortured most brutally and shot dead in the presence of the people.

A few years later – in 1953 – the same treatment was meted out to an ethnic minority group when the state sent its armed forces to Nagaland to suppress the Naga demand for independence. Since then, attempts at self-assertion by other minority groups (e.g. the Sikhs in Punjab, the Muslims in Kashmir) have invited mounting state reprisals. The core of the conflict between these various groups on the one hand, and the Centre on the other, basically concerns how to find an amicable solution to relations between the central authority and the different regional communities that inhabit a multi-national state. Failure to find such a solution within the present constitutional framework of the Indian Union has generated the conflicts. They have been aggravated by the stubborn refusal of the Indian state to acknowledge the failure and by its belief that the conflicts can be solved through military suppression. We must keep in mind the various dimensions of this basic contradiction between a centralizing authority represented by the Indian state on the one hand, and, on the other, the independent aspirations of the numerous regional, linguistic, ethnic communities who are supposed to constitute an Indian nation.

Let us come back to the issue of built-in violence in the structure of the Indian state. A repressive apparatus consisting of the police and the army, which the state inherited from the former British colonial rulers, has been augmented throughout the years by increasing inputs of militarization. In 1949, it raised the CRPF (Central Reserve Police Force), modelled on the British government's plan of Crown Representatives Police Force. Since then, the number of central paramilitary forces has increased steadily, with the formation of the BSF (Border Security Force), the Assam Rifles, the Indo-Tibetan Border Police, the CISF (Central Industrial Security Force), the National Security Guard and the commando force called 'Black Cats'.

Most of these paramilitary forces have been used all these years – and still continue to be used – to tackle civil disturbances within India. The CRPF and the BSF, for instance, were deployed to suppress the Naxalite peasant uprisings in West Bengal (1970–71) – and are being used in Kashmir today to quell the secessionist movement there. Reports of civil liberties groups suggest that military actions in Kashmir are marked by indiscriminate arrests and torture of innocent citizens not involved in the secessionist movement. Nor is this confined to Kashmir. In any part of India, whenever the state deploys its police or paramilitary forces, or the army, it is the innocent citizens who suffer the most – a fact brought out by documentary evidence through on-the-spot investigations by numerous human rights groups in India.[3]

The roots of such arbitrary state violence against ordinary citizens lie deep in the attitude of the Indian state towards popular grievances. The pattern of state response to such grievances follows a common sequence of policy decisions and actions – whether in relation to demands of poor peasants, industrial workers or ethnic minorities. At the initial stage of any demonstration of popular demands, the state then decides to ignore them. When accumulation of the ignored grievances manifests itself in desperate militant agitation, the state decides to treat them as law-and-order problems and deploys its police to suppress them. Such deployment often helps the state to contain the outbursts temporarily and prevent them from exploding into armed insurgencies in what is known as the 'heartland', i.e. the central areas of India. But it is a different situation in the border states in the north and the north-east. While the Indian state has managed to contain the Naxalite movement in Andhra Pradesh and Bihar within the confines of sporadic and brief armed encounters, it has to reckon with better-armed and better-organized insurgencies in Punjab, Kashmir, Manipur and Assam – all border states.

There is of course a basic difference between the political character of the Naxalite armed movement and that of the insurgencies in Punjab, Kashmir and the north-east. While the former seeks to base itself on the politics of class-conflict, the latter are concerned with more general issues and concerns related to their respective communities, irrespective of the class differences that might otherwise divide them. They are trying to bring to the fore the sectarian politics of community-based shared experiences and beliefs (whether based on the religious identity of the Sikhs in Punjab and Muslims in Kashmir, or the linguistic identity of the Assamese in the north-east, or the tribal identity of the Nagas and Mizos). Attempts to bind together members of each community through their respective historical identities frequently prevail over attempts to forge a unity of the depressed sections among all these communities with the common goal of ending class differences.

Apart from this advantage of community-based sympathies that the secessionist movements command in their respective areas in the Indian border states, in terms of military tactics they also enjoy benefits of a military catch-ment. In the north-west, Pakistan is said to provide direct aid to Khalistani (Sikh) and Kashmiri secessionists, while in the north-east the no-man's land in the bordering areas of Burma has become an ideal shelter for the insurgents of Nagaland, Manipur and Assam.

5.4   Growth of Terrorism

The most demonstrative response to state repression in Punjab and Kashmir has been in the form of 'terrorism' as distinct from organized mass move-ments, or guerrilla warfare. Guerrilla-based insurrections in China, Vietnam and Latin American countries in the past involved the willing participation of the local people, both in direct and indirect ways, thanks to a leadership wed-ded to a democratic and socialist ideology which took their people in confidence, and anchored their ideology in popular nationalist aspirations.

Unlike those forms of armed struggle, the insurgencies in Punjab and Kashmir are dominated by a leadership which aims at harnessing popular nationalistic urges to the objective of establishing theocratic autocracies in these states – as evident from the published statements and public utter-ances of the Khalistani leaders of Punjab as well as of the various militant secessionist groups of Kashmir. They appear to owe allegiance to a form of religious fundamentalism (of the Sikh variety in Punjab and its Islamic coun-terpart in Kashmir) that aims at subjugating members of their respective communities to conservative norms and practices, selectively excerpted from their traditional religious scriptures and systems on the plea of restoring the 'fundamental' purity of their respective religions. While imposing such tra-ditional norms on members of their particular communities (like the injunction of the veil by Islamic fundamentalist militant groups on Kashmiri Muslim women who till recently had remained free from such obligations) has threatened the democratic rights of their own people, the professed objective of establishing theocratic states based on strict adherence to their religious doctrines and customs has threatened to force open fissures between Muslims and Hindus in Kashmir and between Hindus and Sikhs in Punjab, where Khalistani militants have launched selective assaults on Hindus.

Insurgencies in the two states enjoy various degrees of popular support from their respective religious communities – prompted at times by admira-tion, at other times by fear – ranging from those who see terrorism as mere retaliation against the state's oppression, to those who support it as a means of achieving independence. But, unlike the national liberation movements in Vietnam or South Africa, authoritarian ideology and distrust in mass mobi-lization have led the leaders of secessionist movements in Punjab and Kashmir to rely almost exclusively on individual or group terrorist actions. Since terrorism without publicity is a weapon that fires only blanks, such ter-rorist actions have had to be directed not only against representatives of the state (policemen, security forces and government officials), but also at gatherings of ordinary innocent citizens. Bomb blasts in crowded places that kill indiscriminately a large number of people invariably become news and give the terrorists maximum publicity, as happens almost every day in Punjab. Since their ideology is intolerant of dissident views, they also attack those members of their own communities who refuse to accept their fundamentalist religious views. In Punjab, Khalistani terrorists have killed a great many leaders and cadres of Leftist parties who dared to oppose their fundamentalism. In Kashmir again, some of the victims of terrorist violence have been veteran Muslim leaders with liberal leanings like Mirwaiz Maulana Farooq and Maulana Masoodi. Reliance on weapons to the exclusion of politics has also led to the criminalization of these movements in a large measure – with the infiltration of the terrorist ranks by gun-runners and smugglers, hired killers and goons out to settle their personal scores.

But paradoxically enough, one of the contributing factors that has swelled the ranks of terrorists in Punjab and Kashmir is the very nature of the state's anti-terrorist operations. Such operations, as noted earlier, invariably direct themselves against the easily visible ordinary citizens instead of the invisible armed insurgents. Thus, the innocent victims of state repression become increasingly alienated from an unpopular administration; many among them – particularly the youth – go underground and join the insurgents, either for protection or for vengeance. According to the Indian government's own admission, the number of terrorists both in Punjab and Kashmir has risen in the last few years. And this despite the deployment of paramilitary forces in increased strength in these two states. Temporary successes like the 'cleaning up' of the Golden Temple in 'Operation Bluestar' in 1984, or the arrest and killing of a few 'terrorist' leaders in Punjab, or the claim of flushing out 'terrorists' by burning entire villages in Kashmir have left a trail of destruction, in the furrows of which a new generation of insurgents is born.

The situation in Kashmir is a classic example. A once-peaceful people have been forced to take to arms, thanks to the Indian state's refusal to accommodate their demands through democratic avenues. The history of the state since its accession to India in 1947 is one of throttling the normal democratic process there through a succession of rigged elections and impositions of unpopular Chief Ministers by New Delhi. The 1987 elections, marked by widespread incidents of rigging, were perhaps the last straw, driving a large number of Kashmiri youth to give up hope in the democratic process and to join the insurgency movement, demanding independence from the Indian Union. Instead of sympathetically heeding their grievances, the Indian state as usual responded with a military crackdown which invited counter-violence from the insurgents. Thus, the cycle of violence continues.

Occasionally, such cycles are interrupted by the signing of 'accords'. The state, following the old colonial policy of 'divide and rule', selects sections of the insurgents and decides to negotiate with them. In 1975, it came to an understanding with some members of the underground Naga National

Council through the Shillong Accord. The rebels were offered amnesty in exchange for surrendering their weapons and accepting the Indian Constitution. But this did not solve the problem of Naga insurgency. Other members of the Naga National Council broke away to form the NSCN (Nationalist Socialist Council of Nagaland), which is still carrying on armed resistance against the Indian state, in pursuit of its goal of an independent Nagaland.

Following the same approach, the government headed by Chandrashekhar sent overtures to the insurgents in Punjab and Kashmir for talks in 1991. Preliminary talks were started with a Sikh leader, Simranjit Singh Mann, who claimed to represent the militant outfits of Punjab. But several militant groups immediately disowned him. Thus, there does not seem to be any likelihood of an end to the cycle of violence that has enveloped Punjab.

The history of the Naga insurgency and the increasing spread of terrorism in Punjab and Kashmir in recent years have proved the resilience of such violent trends even in the face of powerful state offensives. Ironically, while Mao's followers in India – the Naxalites – have so far generally failed to implement successfully his motto, 'Power flows from the barrel of the gun', the truth of his saying is being proved every day in Punjab by religious fundamentalist Khalistanis. At gunpoint they can force government officials to toe their line, terrorize the common people to support them, extort money from the rich to buy more weapons, impose their fundamentalist norms on the citizenry (like making women give up wearing jeans, or keep their heads covered, or forcing the menfolk to wear turbans and grow beards) – in short, be able to run their writ across vast stretches along the border.

Organized terrorism in India, as elsewhere, has been able to acquire newer weapons in an age when sophisticated technology is not contained by national boundaries. The terrorist outfits have kept pace with the Indian state in the arms race. While the state's armed forces compete with their Pakistani counterparts by bidding for the latest weaponry, bombers and submarines from the West, the terrorists have been buying from the same source armaments that suit their own military tactics. They have graduated over the years. The weapons used by the assassins of Indira Gandhi in October 1984 were the vintage .35 pistol and a Thompson submachine carbine (both ironically, acquired from the arsenal of the state's security forces). In May 1991, Rajiv Gandhi was killed by an explosive device so sophisticated that Indian sleuths were still trying to understand the mechanism three years later.

The language of the gun is thus becoming a decisive force in political discourse in India. It is gradually edging out debates in a democratic framework, and suppressing dissent of the traditional humanitarian variety. Both the Indian state and its armed opponents in Punjab, Kashmir and elsewhere refuse to provide any neutral space for expression of disagreements. This trend of intolerance of democratic dissent was inaugurated in Indian politics by the post-Independence state in general, and in particular by the Congress Party which has run it for most of the time since 1947.

5.5   Erosion of Democratic Norms and the Culture of Violence

The Indian state set the precedent of violating democratic norms when, in 1959, Nehru's central government dismissed the first democratically elected Communist ministry in Kerala. Nehru's own Congress Party ganged up behind upper-class, privileged communities there who mounted an agitation against the efforts of the Communist ministry's government to change the status quo in education and agriculture, thus threatening the vested interests of these communities. Since then, the Congress-run Centre has dismissed several non-Congress state governments on one pretext or another.

At the level of political relations between the ruling party and the opposition, the Congress, and later its variant the Congress (I), replaced the language of democratic debate with that of violent confrontation. This became brazen-faced during the prime ministership of Indira Gandhi, particularly during the Emergency Period, when her party goons under the leadership of her son, Sanjay Gandhi, went around snuffing out any spark of dissent. Congress musclemen virtually dictated the law in those days.[4] The Congress culture of violence manifested itself in the most murderous way in November 1984, when following the assassination of Indira Gandhi, her party-men organized a full-scale massacre of 3,000 Sikhs in the streets of Delhi. Both human rights groups and affidavits submitted by survivors of the massacre named leading Congressmen like H.K.L. Bhagat and Sajjan Kumar as responsible for organizing the killings.[5] To date, none of the accused Congressmen has been prosecuted. Even worse, the Congress (I) nominated these same persons as candidates in the 1991 parliamentary elections.

Thus, while on the one hand the Indian state fails to punish those guilty of mass killing of members of the Sikh community, on the other hand it indulges in violent reprisals against the Sikhs in Punjab on the official pretext of 'suppressing terrorism'. The country's existing penal laws are waived for Congressmen guilty of heinous crimes while new draconian anti-'terrorism' laws are being passed which deny citizens the fundamental rights guaranteed under the Constitution. The TADA (Terrorist and Disruptive Activities Prevention Act) provides for preventive detention and withholds minimum safeguards for fair trial. Under this Act, the trial of the accused is held *in camera*, with the onus of the proof of guilt shifted from the prosecution to the accused, whose extrajudicial confession is deemed good enough for admissibility in the courts. The NSA (National Security Act) empowers the authorities with the right to refuse to disclose the facts relating to grounds of detention and to prevent the detainee from having a lawyer to represent his/her case. Under the Armed Forces Special Powers Act, armed personnel of the Indian state enjoy the right to fire upon anyone and kill, without any accountability whatsoever. The law provides them with immunity against prosecution. These are only a few among the many undemocratic laws currently in operation in Punjab, Kashmir and the north-eastern states.

Ostensibly meant to curb terrorism, these laws contain provisions so far-reaching and arbitrary that they act to legitimize the indiscriminate persecution of ordinary citizens by the security forces. The contrast between the state's refusal to prosecute criminals against whom there is *prima facie* evidence but who enjoy political protection on the one hand, and the state's proclivity towards harassment of innocent citizens who do not enjoy such protection on the other, was expressed bluntly by the Sikh leader Simranjit Singh Mann in his memorandum to the then Prime Minister Chandrashekhar (28 December 1990):

> . . . while in Punjab people can be shot by the police on the vaguest suspicion without any process of law, those guilty of killing Sikhs in 1984 freely roam the streets of Delhi.

The discriminatory treatment is rooted in the policies of the ruling elite which are heavily loaded in favour of those who are powerful, and which discriminate against those who are not. In an unequal society, the coercive role of the state in its violent manifestation betrays the same discriminatory tendency, and as a result invites a violent response from those who suffer discrimination.

It is against this background of state repression and counter-violence that we should look at the assassination of Rajiv Gandhi in May 1991. If he fell victim to a much talked-about 'remote device' detonator, we must also remember that the original politically explosive 'remote device', metaphorically speaking, had been planted in Sri Lanka back in 1987 by the IPKF (Indian Peace Keeping Force) sent there by the Indian state, which set off chain reactions from which both Colombo and New Delhi have yet to recover.

But apart from the controversy over the motives of those who killed Rajiv Gandhi, the assassination highlights the general culture of violence that marks Indian society today. It should not be viewed as a sudden isolated exception, though no one denies that this particular event was to have political implications more far-reaching than those generated by the killings of ordinary Indian citizens in political violence. More than a hundred citizens – political workers as well as innocent bystanders – lost their lives in India in the course of the fortnight from 14 to 28 May 1991. The first week of the period preceded Rajiv's assassination, when those killed became victims of violence related to the electoral campaign on the eve of the 1991 general elections to the Indian Parliament and state legislatures. The second week, following the assassination, saw widespread attacks by Congress (I) activists on the lives of members and sympathizers of their opponent parties, whom they chose as targets for mindless vengeance for the loss of their leader. Violence has indeed become the dominant medium of expression – an expression of political loyalties during a democratic exercise like elections as well as an expression of emotional outbursts at the death of a leader.

5.6 Electoral Violence and Communal Violence

Secessionist terrorism and insurgencies, on which we have concentrated so far, can be described as violent demonstrations of self-assertion by regional groups (often based on sectarian religious ideologies as in Kashmir and Punjab) against violent manifestations of centralization by the state. But two other types of violence – the one erupting during elections, and the other around Hindu–Muslim relations – stem from acts of omission by the same state.

The escalation of violence in these two areas in recent decades indicates yet another contradiction from which the Indian state suffers. On the one hand, the pillar of 'democracy' on which the Indian Constitution is built allows popular participation in the political system within a framework of rules, rights, structures and processes; thus, such participation acquires a certain legitimacy. But on the other hand, at the micro-level, popular attempts at participation are thwarted by powerful vested interests. Particularly during elections, in the countryside, members of the downtrodden, depressed lower castes are often prevented from voting by the upper-caste landlords. Over the years, muscle power has emerged as the decisive force in elections. Whoever can employ goons to terrorize the voters, capture polling booths and mark ballot papers in his favour, stands a better chance of winning the elections. The extent of such rigging can be judged by the fact that after the first phase of polling in the 1991 elections, India's Election Commission had to order repolling in at least 1,000 polling stations in seven states because of violence which had disrupted the pursuit of fair elections.

Similarly, the other Constitutional pillar, that of 'secularism', is tottering under increasing communal tensions in Indian society. Although the state swears by 'secularism', it has done little indeed to secularize a civil society that remains extremely backward, ridden by caste prejudices and religious intolerance, superstitious beliefs and obscurantist practices. In Hindu–Muslim relations in particular, the pre-Partition legacy of communal division has been allowed to intrude into politics, lending legitimacy to both Hindu and Muslim fundamentalism. In the general environment of violence, communal politics also is assuming the form of violent confrontation between Hindus and Muslims. The year 1990 marked an alarming rise in the graph of Hindu–Muslim riots that had been increasing steadily all over India for several years. The strident campaign by the Hindu communal forces to demolish the Babri Masjid, the Muslim place of worship, in Ayodhya (north India) and build a temple there led to widespread disturbances where most of the victims were Muslims.

The subversion of democracy in elections, and of secularism in civil society through violent means, has been made possible largely by what is commonly called the 'criminalization of politics'. A vast underworld network of smugglers, hired killers, gun runners and gangsters, thriving over the years on the economy of 'black money', has now acquired legitimacy and power in society, thanks to the patronage and protection it receives from politicians who use members of this underworld to terrorize their opponents and impose their control over the people. This underworld plays a decisive role in rigging elections and fanning the flames of communal riots.

5.7   Conclusion

The roots of violence in Indian society today are historical and lie in the contradictions inherited from the pre-Independence era. These contradictions express themselves in various forms, ranging from constitutional debates between the Centre and the state governments over the question of reallocation of resources, to extra-constitutional armed confrontation between the ruling authorities and secessionists demanding independence as in Punjab, Kashmir and the north-east. These contradictions spill over into multilevel violent feuds, like class conflicts between landlords and landless peasants, fratricidal warfare between one caste and another, and communal riots between Hindus and Muslims.

In the incomplete process of nation-building, the Indian state is yet to overcome these unresolved contradictions between traditional regional, linguistic, ethnic, religious and other identities which divide the Indian people and impede their struggle to establish an egalitarian society.

In this situation, the Indian polity has two options. *One*, it can evolve a framework to harmonize these different identities that divide society today, at the same time allowing the peaceful development of each group. Decentralization of power and attempts at an equitable reallocation of resources could be the guiding principles for evolving such a framework. This would require a change in the constitutional structure, which Indian politicians must realize is not something that is unalterable. One such change was proposed by the Sikh political organization Shiromani Akali Dal in October 1973 and has come to be known as the Anandpur Sahib resolution. A source of intense controversy, the resolution can be stripped of its religious overtones and reduced to its basic demand – that the role of the Central government should be confined to defence, currency, foreign relations and communications, while the states should be allowed to take care of the other subjects.

There is no harm in opening up this demand to a national debate, inviting participation from all sections of the Indian people – including representatives of national political parties, central civil services and monopoly industrial houses (whose interests are pan-Indian) as well as those from the regional political parties, agrarian lobbies of different class interests, and the numerous non-party people's movements who are demanding protection of the environment, tribal culture and habitat, and women's rights and are engaged in training villagers in developing a scientific outlook. Dialogues can be initiated with those outside the mainstream of constitutional politics – the secessionist groups in the Punjab, Kashmir and the north-east, as well as the various armed Naxalite groups which are fighting a class war – on the possibilities of changing the structure of the Indian constitutional system.

A national debate and dialogues at various levels might lead to the emergence of a new model of development for Indian society on the basis of better participation of the people. Admittedly this could not be a model based on consensus, given the class differences and conflicts that dominate the country's economy. But it might help in resolving some of the social contradictions within a democratic framework, however time-consuming the transformation might be in its evolutionary way. Attempts at restoring and reinforcing the democratic structures – as well as building up new ones – will, in the future model of development, have to be accompanied by determined administrative efforts to eliminate the sources of 'criminalization of politics', so that the language of violence can be replaced by that of democratic debates.

The option available to Indian politicians as outlined above can by no stretch of the imagination be described as an attempt at a revolutionary transformation of Indian society. At best, it is an effort to make some structural changes in the administration that can allow increased democratic participation of the people, without any radical changes in the basic economic system. But is today's Indian polity innovative enough to bring about even these structural changes?

The *other* option remaining is a passive witnessing of a slow and disastrous disintegration of Indian society presided over by a state stubbornly dedicated to the doctrine of centralization through repression. In an atavistic return to its violent past, Indian society may look for a solution to its conflicts in blood-letting – the same bloodbath that once led to the partition of the sub-continent and brought the Indian state into being.

Notes

1. There is voluminous research work on Hindu–Muslim communal relations in colonial India, important among which are Mushirul Hasan (1979) and D. Page (1981). For class conflicts in rural India, see A.R. Desai (1979). Caste conflicts have been covered in Gail Omvedt (1976) and Lloyd I. Rudolph & Susanne Hober Rudolph (1967). Of the literature on ethnic conflicts in the north-east, mention can be made of Neville Maxwell (1973).

2. This history of Kashmir politics since the accession is fully documented in Balraj Puri (1981).

3. A collection of these investigative reports is to be found in A.R. Desai (1990).

4. Evidence of such hooliganism was submitted by victims to the inquiry commission headed by retired Supreme Court Judge F.C. Shah and set up by the then Janata government at the Centre in 1977 (after the lifting of the Emergency and the defeat of Mrs Gandhi in the elections that year).

5. See *Who are the Guilty?* (People's Union for Democratic Rights . . ., 1984).

References

Bose, Nirmal Kumar, 1962. *Studies in Gandhiism*. Calcutta: Merit Publishers.

Desai, A.R., 1979. *Peasant Struggles in India*. Delhi: Oxford University Press.

Desai, A.R., 1990. *Repression and Resistance in India*. Bombay: Popular Prakashan.

Hasan, Mushirul, 1979. *Nationalism and Communal Politics in India 1916–28*. Delhi: Manohar Publications.

Maxwell, Neville, 1973. *India and the Nagas*. London: Minority Rights Group.

Omvedt, Gail, 1976. *Cultural Revolt in Colonial Society: the Non-Brahman Movement in Western India, 1873–1930*. Bombay: Scientific Society Education Trust.

Page, D., 1981. *Prelude to Partition: All-India Muslim Politics, 1921–32*. Delhi: Oxford University Press.

People's Union for Democratic Rights and People's Union for Civil Liberties, 1984. *Who are the Guilty? Report of a joint enquiry into the causes and impact of the riots in Delhi from 31 October to 10 November*. Delhi: PUDR & PUCL.

Prasad, Pradhan H., 1989. *Lopsided Growth: Political Economy of Indian Development*. Bombay: Oxford University Press.

Puri, Balraj, 1981. *Jammu and Kashmir – Triumph and Tragedy of Indian Federalism*. Delhi: Sterling Publishers.

Reddy, Ravi Narayan, 1973. *Heroic Telengana*. Delhi: People's Publishing House.

Rikhye, Ravi, 1990. *The Militarization of Mother India*. Delhi: Chankya Publications.

Rudolph, Lloyd I. & Susanne H. Rudolph, 1967. *The Modernity of Tradition*. Chicago, IL: University of Chicago Press.

6

Militarism and the Militarization of Pakistan's Civil Society: 1977–1990

SHIREEN M. MAZARI

6.1 Introduction

Whether it is a conflict between hostile student organizations, or political rivalry, or an argument between landlord and tenant, Pakistan's civil society has seen an increasing resort to violence to resolve all manner of conflicts. This militarization of civil society – its preference for a violent course of action over other means of exercising influence – became marked after the imposition of martial law following a military coup in July 1977. Therefore it is pertinent to examine the development of militarization within state and society in Pakistan – within the framework of the concept of militarism which denotes a 'social formation and structure' (Thee, 1980; Wallensteen et al., 1985).

In order to establish a clear concept of civil society, we will use the Gramscian distinction between state (in the narrow sense) and civil society. The 'state' refers to the coercive elements of government – that is, the politico-juridical organization in a narrow sense – while 'civil society' refers to a 'multiplicity of private associations (of two kinds: natural and contractual or voluntary)' such as churches, trade unions and political parties, to name a few (Gramsci, 1971, pp. 261–264).

Militarism has not only expanded at the international level, in the form of arms buildup and in the use of force as an 'instrument of supremacy'; in the developing states it has also led to an increase in the role of the military establishment in domestic and foreign policy-making and execution. As the military profession has developed massive defence budgets, it has established greater autonomy and assumed a privileged position in society – especially in states where civil institutions are weak and dissipated.

In such situations, the military evolves its own doctrines of state and government,[1] which drive it to seek to influence state formation. This sense of purpose, along with the desire to increase its power and privilege within the state, leads the military to abuse its legitimate function and make inroads into the political structures of the state. Thus, in several developing states the military exercises a decisive influence on state policy, either by directly taking over the structures of the government or indirectly by controlling and/or manipulating the civilian ruling elite. This leads to the state, in the broad sense (government and civil society), being run by a military-bureaucratic-corporate-intelligentsia (MBCI), with the military predominating (Wallensteen et al., 1985, p. *xii*).

While militarism seeks legitimacy, in general it relies on force to perpetuate itself. As Pakistan's General Zia stated, 'martial law should be based on fear' (Noman, 1988, p. 22). The use of force by the MBCI alliance becomes more pronounced where civil society as a whole, or critical segments within it, reject militarism. This leads to increasingly violent behaviour, with militarization pervading the state and civil society as a preferred means of exercising influence. Therefore, militarization is directly linked to the concept of militarism – reflecting it at the behavioural level of state and civil society. Both militarization and militarism also reflect the prevalence of a conflictual framework at the level of the state and civil society, where increasing violence comes to mark conflict behaviour – not only of the state but also of civil society within the state.

Marek Thee has identified a variety of forms in which militarism can exist, ranging from a repressive authoritarian regime backed by the military, to direct military rule, to civilian rule with the military exerting predominant influence (Thee, 1980, pp. 19–20). Depending upon the prevalent form, differing degrees of violence within civil society accompany this militarism. To evaluate the form and extent of militarism and militarization it is useful to refer to indicators identified by Thee, which fall into three categories.

The first set of indicators relates to the structure of the government – Thee terms these the 'structural systemic features' – and are primarily derived from: (a) the position of the military in the state system – whether it is the ruling force, or it is in co-partnership with the civilians or whether its decisive authority is under overall civilian supremacy; and (b) the extent of deviation from democratic rule – ranging from dictatorship, to denial of democratic freedoms, to repressive measures (Thee, 1980, pp. 19–20).

The second set of indicators relates to the ideological strands of militarism. Amongst the 'elements of ideology', Thee refers to notions ranging from nationalism to xenophobia (1980, pp. 19–20). Militarism also contains within it elements of aggressiveness and bellicosity, and often focuses on ideological dogmatism. There is also a stress on glorifying the military establishment and emphasizing hierarchy, discipline and regimentation.

The third set of indicators relates to the focus of state policies and the form of their execution (Thee, 1980, p. 21). These can range from high military expenditures and policies of preferential treatment to the armed forces, to an inclination towards using military strength as an instrument of politics and diplomacy. Militarism also indicates the military interfering in the country's internal security operations. Thee asserts that militarism often reflects right-wing political orientations and a preference among the ruling elites for participation in military alliances.

6.2   Pakistan: Background

Within the above framework, militarism has prevailed in Pakistan for most of its history since the first imposition of martial law in 1958. Whether directly or indirectly, the military has therefore exercised a decisive influence upon Pakistan's domestic and foreign policies for over three decades. In order to understand developments in the 1977–90 period, it is essential to understand the events immediately following the creation of Bangladesh and the temporary withdrawal of the army from politics.

The military defeat by India and the loss of East Pakistan in 1971 led to a short period during which militarism abated within the Pakistani state, with the military demoralized and its organization weakened – not only as a result of defeat but also because of the 90,000 prisoners of war being held by India. Its loss of credibility within the nation reinforced the institutional demoralization of the army – the main component of Pakistan's Armed Forces.

Zulfiqar Ali Bhutto's assumption of the Presidency in December 1971 – he became Prime Minister with the promulgation of the 1973 Constitution – led to a brief period when several measures were taken by the civilian ruling elite to reassert its supremacy over the military.

First: Bhutto appointed a commission of inquiry headed by the Chief Justice of Pakistan, Justice Hamood-ur-Rahman, to investigate the circumstances that had led to the military disaster in East Pakistan and the acceptance of a ceasefire in West Pakistan.

Secondly: several senior military officials, all of whom had been associated with General Yahya's regime and the conduct of the 1971 war, were removed from their posts (Askari, 1987).

Thirdly: changes were made in the administrative setup of the military high command, with the designation of the heads of the three services being altered from Commander-in-Chief, to Chief of each relevant service. The President of the country became the sole Commander-in-Chief. The three service chiefs were also put under the command of the Joint Chiefs of Staff Committee to develop a greater degree of professional cooperation and policy coordination.

Fourthly: the tenure of the chiefs of the three services was initially fixed for four years; in 1975 it was reduced to three years. It was also decided not to grant extensions to the services' chiefs, in order to prevent them from being able to consolidate their hold over their forces (Askari, 1987).

Fifthly: for the first time, the functions of the military were defined within the Constitution. Article 245 of the 1973 Constitution states that the military, under the direction of the Federal Government, is required to 'defend Pakistan against external aggression or threat of war and, subject to law, act in aid of civil power when called upon to do so'. The notion of High Treason was also defined very comprehensively in Article 6: 3 (Askari, 1987).

Despite all these efforts, the Bhutto government soon found itself facing an increasingly militarized civil society, with intra-provincial and ethnic conflicts erupting in the aftermath of the events of 1971. The political situation was further aggravated by the Bhutto government's inability to accept non-PPP (Bhutto's Pakistan People's Party) governments in two of the provinces (Baluchistan and Frontier Province). At the same time the multiple strands of ethnic tensions in Sindh – between the Urdu-speaking settlers and indigenous Sindhis on the one hand and the Punjabi settlers and Sindhis on the other, for example – were beginning to surface in a violent fashion.

Bhutto therefore felt a need to rebuild and strengthen the army while limiting its powers in the Constitution. Thus, the Bhutto government allocated more resources to the military than any previous government – with defence expenditure showing a rise of around 89% during 1971–72 and 1975–76, and including increases in salaries and other allowances (Askari, 1987). At the same time the army was called out by the civil government to maintain law and order on several occasions between 1972 and 1977.

By the end of 1972 the initial signs of an insurgency movement were visible in the province of Baluchistan, and by 1973 the army had become overtly involved in fighting this insurgency spearheaded by two of the main Baluch tribes – the Marris and the Mengals. The army was also called out to deal with language riots in Sindh Province in July 1972, anti-Ahmadya riots in June 1974 and the conflict between the civil administration and tribesmen of Dir, in Frontier Province, in October 1976.

The politicization of the military was reflected in growing discontent within the army and air force over government policies; this resentment surfaced as early as 1973 when 14 officers of the air force and 21 army officers were arrested on charges of conspiring to promote a coup d'etat. One reason for this resentment was the sheer frustration within the forces over what was perceived as a weapons lag in relation to India, especially with the cut-off of US weapons supplies. Bhutto's socialist and Third World orientation was seen as partially responsible for the continuing US hostility towards Pakistan. The issue was further aggravated when Pakistan's nuclear policy led to an open rift with Washington. Also, according to some analysts, the military felt a general rejection of mass politics (Pinkney, 1990, p. 65).

Although the Bhutto government had increased the defence budget and in 1973 had set up a Defence Production Division in the Ministry of Defence to encourage indigenous production of arms and ammunition, the military remained dissatisfied. Therefore, the 1977 street agitation launched by a coalition of political forces – calling itself the Pakistan National Alliance – against the alleged electoral rigging by the Bhutto government was used by the army to impose the third martial law regime in Pakistan.

6.3 Pakistan 1977–1990

6.3.1 The Politico-ideological Framework of the Military Regime

The 1977 coup reflected the resurgence of military power within Pakistan. The strength of the Pakistan Army within the country and its organizational recovery, despite the 1971 military defeat and the politicization after long years of being involved in governing the country, lay in its size – almost half a million strong – and its professionalism. In Pakistan, especially after 1971, the military's professionalism had developed alongside its politicization. Even during periods of martial law, the military as an institution did not get directly involved in the civil administration, but exerted control through a military-bureaucratic alliance. Equally critical was the fact that the military defeat in 1971 led to a whole process of restructuring within the army, with emphasis on the education of army officers – including sending them to foreign universities.

The 1971 experience also shifted the military's politicization within a more ideologically motivated framework which extended beyond the military's traditional focus on the national interest and its own corporate identity. Akmal Hussain points out that after 1971, the officer corps was exposed to indoctrination by the Jamaat-e-Islami, the party of the religious right, and the left wing of the PPP (Hussain, 1989, p. 218; see also Jalal, 1990, pp. 317–324). Not only were a growing number of officers products of the indigenous educational system, where they had been exposed to systematic political indoctrination – especially by the Jamaat-e-Islami (JI), which dominates most campuses – in addition, the JI had over the years made a concerted effort to penetrate the officer corps with its own cadres. A number of young officers were also influenced by Bhutto's socialist rhetoric and saw in him the 'harbinger of a strong new Pakistan' (Hussain, 1989).

With the prevalence of politicization, by both right and left, the army lost its neutrality and thereby its ability to act independently of any particular political group. Thus, when General Zia's coup took place, it did not create a veneer of ideological neutrality. Instead, General Zia could draw on the Islamic ideology theme used by the PNA movement against Bhutto and concentrate on winning over those social groups which had formed the main opposition to Bhutto – the small shopkeepers and petty merchants of the lower middle classes, as well as the industrialists and big businesses affected by Bhutto's nationalization. He also utilized the *biradri* (kinship) system; so, unlike under previous martial laws, the military regime's efforts to maintain itself in power extended beyond structures of state power into the dynamics of social relations within civil society. This meant that the cleavages within Pakistan's civil society were accentuated by the Zia regime, as it attempted deliberately to win the support of segments of civil society rather than civil society as a whole.

Initially, the coup had been justified by the claim that Pakistan had been on the verge of a civil war. But when that failed to convince people and their weariness with military rule became apparent,[2] General Zia began to feel the need to legitimize his rule. He therefore talked in terms of holding the 1973 Constitution in abeyance rather than abrogating it, and claimed to be working towards creating structures for an Islamic state and society within Pakistan. Yet, by allying with particular segments within civil society, the efforts of the Zia regime to legitimize its rule tended towards further polarization within Pakistan's civil society.

Zia's efforts at Islamization came to be linked to the precepts of a particular brand of Islam (that of the Jamaat-e-Islami) and led to sectarianism, with religion becoming enmeshed within political controversies. Most efforts at Islamization were aimed at women, with a series of discriminatory ordinances being promulgated. For example, there were the Hadood Ordinances (1979) which made adultery indistinguishable from rape, and the proposed Qisas and Diyat Ordinances through which 'blood money' for a woman was half that of a man. The Law of Evidence (1984) also led to the evidence of a woman witness being equivalent to only half that of a man.

While these aroused the opposition of some women, protests were initially largely confined to the Western-educated urban elite. It was Zia's efforts to Islamize the economy that divided society along sectarian lines, with the Shias vociferously protesting against Zia's attempt to impose *ushr* (agricultural tax) and *zakat* (alms tax). Although Zia exempted Shias from these two taxes, for the first time in Pakistan the population had to identify itself officially along sectarian lines. From this point, all the regime's efforts at Islamization were seen as pushing the interests of a particular sect.

Meanwhile, in the face of political opposition to military rule and in order to forestall any mass mobilization in support of the deposed leadership, the military overtly and deliberately used psychological warfare against civil society by generating fear. While the international community tended to associate Zia's public floggings and hangings with the process of imposing his interpretation of Islamization on the country, these measures were in fact announced *before* his Islamization programme had commenced. A few days after the coup, on 10 July 1977, military courts with absolute power were set up, and martial law regulations were enforced which instituted public floggings and hangings. According to Noman (1988, p. 122): 'the conscious use of terror as an instrument of domestic policy was implemented through the uninterrupted use of Martial Law between 1977–1985'. The hanging of deposed Prime Minister Bhutto in April 1979 symbolized the violence and terror the military was ready to use to maintain itself in power.

6.3.2   Interaction Between State and Civil Society

The hanging of Bhutto, of course, further widened the schism between the military regime and the people, especially in rural Sindh – Bhutto's home province. The majority of the population in Sindh is Sindhi-speaking, but over 20% speak Urdu and they predominate in the urban areas, being referred to as '*mohajirs*'. While the Sindhis see better-educated *mohajirs*, along with other non-Sindhis like the Punjabis, as exploiters attempting to monopolize jobs and privileges in the province, the *mohajirs* resent the quotas (all government jobs are determined by provincial and urban–rural quotas), which they see as restricting employment opportunities. The 1972 language riots had already shown the potential for violence between the Urdu-speaking and Sindhi-speaking population in Sindh (Alavi, 1989).

Bhutto's hanging not only fed the Sindhi/non-Sindhi conflict, but also created overt hostility between the military and the Sindhi population. The latter saw the military increasingly as a 'Punjabi' force acting against the interests of Sindh, and the power of the central government became identified with a particular ethnic group. The policies of General Zia's regime had led to the PPP becoming the symbol of Sindhi nationalism. According to one report, the ruthless suppression of Sindhis in general, and the PPP workers in particular, served to fuse nationalist and pro-PPP sentiments (*Herald*, January 1988, pp. 45–54).

In order to undermine political support for the PPP in Sindh, the Zia regime had seemed to encourage the *mohajirs* to come together under one political platform; the first formal assertion of *mohajir* identity had been the creation of the All Pakistan Mohajir Students Organization in 1978. However, the PPP continued to be the single largest political party at the national level; and the 1979 local body elections, although held on a non-party basis, saw the election primarily of candidates affiliated with the PPP. This continuing support for the PPP led General Zia to impose a ban upon political parties, which in turn meant further polarization between the military and the populace.

It was an external factor that gave the Zia regime a new lease of life by allowing the regime to become acceptable within the international community, despite its dictatorial nature and its insistence on hanging Bhutto in the face of concerted pleas worldwide. The Afghan crisis, especially the entry of Soviet troops into Afghanistan in December 1979, made Zia's Pakistan a front-line state in the struggle against this Soviet intervention, and a major recipient of US economic and military assistance. After a gap of over a decade, the Pakistan military was able to support an ambitious modernization programme involving the introduction of some of the latest weapon systems into the forces and the acquisition of advanced military technology. This led to a situation where the professionalism of the Pakistan Army developed alongside its increasingly partisan politicization.

The Afghan crisis also altered qualitatively and quantitatively the militarization of civil society, primarily as a result of two factors:

- There was a massive influx of arms, especially submachine guns and automatic rifles, as weapons meant for Afghan guerrillas proliferated on the illegal arms market (Noman, 1988, p. 200; Hussain, 1989, p. 231).
- There was a rapid growth of the heroin trade with mafia-type syndicates often involving Afghan refugees who controlled the bulk of intercity cargo services (Hussain, 1989).

With political parties banned and all venues for protest through legal means closed, polarization within society intensified. Cleavages and conflicts within civil society, which had already shown a violent trend under Bhutto's increasing use of the coercive elements of the state, grew worse under military rule. The ban on political parties led to an increasing focus on seeking identity through group membership based upon ethnicity, sectarianism and the traditional *biradri* (kinship) system. This further bolstered the prevalent conflict within society, as polarization developed vertically (Hussain, 1989).

The violence that came with the availability of weaponry and the abundance of 'drug' money was often encouraged by the military regime's support of particular parties/groups, and militarization increased as the military shed its political neutrality.

One of the first environments where this factor became clearly visible was on the university campuses, where the Zia regime's 1979 University Ordinance reduced the role of the teacher to a menial government employee and allowed the government to victimize teachers for their political sympathies. Several teachers were either transferred to schools/colleges in the interior of the country, sacked or incarcerated. The University Ordinance was a concerted attempt to destroy the teachers' community which had traditionally resisted government interference on campus.

The traditional hostility between left- and right-wing student organizations was intensified not only with the wholesale introduction of firearms (where previously the main weapons had been brickbats and knives) but with the Zia regime giving sanction to students belonging to the student wing of the Jamaat-e-Islami (the Islami Jamiat-e-Tulaba – IJT). August 1979 saw the wholesale organized use of student firepower on the Karachi University Campus, when the IJT used handguns against a protesting crowd of Progressive Front Students. The occasion was the swearing-in of IJT's Hussain Haqqani as President of the Student Union. After the Soviet intervention in Afghanistan, arms proliferated at a phenomenal pace. Gradually, university hostels were transformed into armed fortresses and student leaders became like urban guerrillas, well-trained in the use of modern weapons.

Nevertheless, it was only after the banning of student unions in 1984 that the conflicts on campuses worsened, since unions had until then remained accountable. Initially, the Zia regime colluded with the IJT. In 1979, the Islamabad IJT spearheaded the burning of the US embassy; even though the leaders were identified, the Zia regime showed reluctance to punish them. However the Zia regime realized it could not control the student unions and hence felt compelled to ban them.

According to one report, while 14 students were killed in incidents of violence on campus between 1977 and 1984, over 80 were killed between 1984 and 1988 (*Herald*, October 1988, p. 52). The ban on student unions also led to the development of ethnic and subregional cleavages, as students began to identify along these divisions instead.

The upsurge in campus violence was but one manifestation of the general political situation in the country. In Sindh, with ethnic and subregional conflicts becoming more violent, the tradition of private armies was revived. The Hur and Sarwari Jamaats (that is, the armed forces comprising the disciples of the Pirs, feudal cum spiritual leaders, of Pagara and Hala respectively) have been in existence for a long time. By the late 1970s, other tribal chiefs followed the Pirs' example and created their own forces – among them were the Magsis, the Unars and the Daharis from Nawabshah (*Herald*, January, 1988).

General Zia's efforts at co-opting some political factions met with only limited success. While individual feudal and business elites were co-opted into his nominated *Majlis-i-Shoora* (Assembly), the mainstream political parties formally remained on the outside. In 1983, all the major political parties of the country launched a collective Movement for the Restoration of Democracy (MRD). Strongest in Sindh, it helped transform the politics of the Province when a new militant force appeared which threatened the feudal component of the MRD as much as the government. The defection of several notable Sindhi feudal politicians to the side of the military regime is attributed to this growing militancy within the Sindh MRD during the 1983–85 period.[3] At the same time, these defections, coupled with the ineffectiveness of the political parties in channelling frustration and disaffection, ultimately led to the creation in Sindh of armed gangs – many of whom went on to join the hordes of dacoits or robber bands.

This led to a strange coalescing of forces where armed political workers linked up with common criminals; it allowed the army to carry out action in rural Sindh, ostensibly against dacoits but in fact resulting in the widespread terrorization of the rural population. In rural Sindh, 1984 saw the start of a sharp deterioration of the law and order situation; in addition to killings, there were, between 1984 and 1986, reports of destruction of property and crops – all at the hands of the government (Human Rights Commission, 1986, 1990).

Dacoity and politics fed ethnic polarization, and in March 1984 the Mohajir Qaumi Mahaz (MQM) came into being. Growing out of the All Pakistan Mohajir Students Organization (APMSO), the MQM was organized along the lines of a fascist party with armed cadres. Amongst the demands of the MQM were recognition as a fifth nationality – alongside the Baluch, Sindhi, Pakhtun and Punjabi – and allotment of a 20% quota at the Federal Government level and 50–60% in Sindh (*Herald*, January, 1988, pp. 85–86).

The MQM arrived on the provincial political scene in Sindh at a time when ethnic tensions were extremely volatile. The MQM aggravated this volatility, leading to the start of regular, organized street violence in the urban areas of Sindh. The first evidence of this came in Karachi in 1985 when a young girl was run over by a Pakhtun bus driver and a clash occurred between the Pakhtuns and the *mohajirs*. In addition to Karachi, other Sindhi urban centres like Hyderabad also faced a breakdown in law and order; between 1985 and 1990 Hyderabad was under curfew 13 times in the aftermath of violent incidents (Human Rights Commission, 1990). For the military, the MQM provided a counter to MRD power in Sindh.

The situation in Sindh became more violent as the military regime increased its repressive measures to undermine the MRD's militancy in the province. People were killed, and many villages were destroyed in rural Sindh. At the same time, General Zia finally managed to co-opt some of the traditional political elite into the system, by successfully holding non-party elections for national and provincial legislatures in 1985. Following the elections, a Sindhi prime minister, Junejo, was nominated and the truncated 1973 Constitution restored, with the new amendments having altered the balance in favour of the President – an office occupied by General Zia.

By 1986 the MRD movement seemed to be petering out, and this led to other changes in Sindh's politics. There was a realignment of political forces as the Sindhi-Baluch-Pakhtun Front (SBPF) and other militant nationalist organizations began to be heard among the masses in Sindh (*Herald*, February, 1986, pp. 40–41).

After 1977, the increasing militarization of civil society in Sindh developed along the lines of ethnic and subregional conflictual frameworks. The drug trade, dacoity, kidnapping and political killings became intertwined, fusing into the ethnic and subregional divides, while readily available weapons increased the spectrum of violence. In such a situation any efforts to maintain law and order became prone to suspicion and retaliation. For instance, the Sohrab Goth 'clean-up' in 1987 in Karachi, which involved the local administration in bulldozing a Pakhtun settlement on suspicion of its being a drug and arms den, led to a series of ethnic massacres and further polarized the ethnic 'battle lines' (Human Rights Commission, 1990).

While Sindh reflected the most extreme case of militarization within Pakistan, the other provinces also saw a growing violence pervade their civil societies. Both Baluchistan and North West Frontier Province, bordering on Afghanistan, saw levels of violence increase in their societies directly as a result of the Afghan crisis and the influx of refugees.

With Pakistan a front-line state in the fight against the Kabul regime, and a conduit for arms to the Mujahideen, the repercussions of the guerrilla warfare became felt in the country, especially in Frontier Province with an increasing number of bombings and acts of sabotage, as well as violent clashes between rival groups of Afghans. In addition, conflicts between local populations and Afghan refugees became violent, as the parties involved were armed.

The linkage between the Mujahideen groups and right-wing, religious parties within Pakistan made the whole Afghan issue part of the domestic political conflicts raging in the country – especially since General Zia had linked up an essentially political issue with religious obligation and a messianic zeal.

In Baluchistan, the influx of Afghan refugees threatened to upset the population balance between the Baluchs and Pakhtuns, and personal disputes often tended to acquire an ethnic colouring and violent character. In 1986, a violent clash in Quetta between groups of transporters led to 12 deaths.

In the Punjab, militarization of civil society was reflected in growing sectarianism and in crimes against women – the latter often at the hands of the coercive elements of state power such as the police. In addition, General Zia's support of the IJT allowed the latter to enforce a reign of terror on the campuses of the Punjab University, with students and teachers compelled to fall in line or be removed.

The sectarian divide became increasingly violent with the creation of militant extremist groupings like the Anjuman-e-Sipah-e-Sahaba (ASS). Formed by a militant anti-Shia *khateeb* (one who leads prayers in the mosque), Maulana Jhangvi, the main objective of the ASS was to undermine the rights of Shias.

The Zia period saw not only the emergence of an increasing number of politico-religious parties into the mainstream of the polity, but also the growing tendency for these parties to use violence against each other – leading to the murder of politico-religious leaders on all sides of the sectarian divide and the subsequent intensification of violent conflict amongst their loyalists.

In May 1988, General Zia's experiment with limited democracy came to an abrupt halt when he dissolved the National and Provincial Assemblies. One underlying reason for this action was what was perceived as a growing independence of the civilian political elite which threatened to betray the military's corporate interest – especially through efforts to move towards a political settlement of the Afghan crisis.[4] More immediately, the dissolution came in the wake of the Ojhri disaster, where an explosion in the Ojhri Ammunition Dump in the midst of the city of Rawalpindi caused vast damage to lives and property, and in the wake of the Junejo government's investigation, which threatened to reveal the involvement of the army hierarchy in the disaster.

Following the dissolution of the Assemblies, the political environment altered rapidly in Pakistan after General Zia and several of the military elite were killed in an air crash in August 1988. The psychological trauma of this act of sabotage against the ruling elite itself led the military to agree to the holding of General Elections on a party basis, in which the PPP emerged as the single largest party. However, the years of Zia's political machinations had their effect, and the PPP failed to gain an overall majority in the national legislature. Benazir Bhutto's prime ministership was therefore the result of a compromise with the existing structures of power, with the division of powers tilted heavily in favour of the President – as in 1985. Meanwhile, the military could gradually withdraw from overt involvement in politics.

Even with the restoration of a civilian democratic government, expectations that militarization would recede from civil society proved futile. As the PPP grew restless within the confines of the prevailing structure of power – a ruling troika composed of the Prime Minister, President and Chief of Army Staff – the ethnic and sectarian conflicts continued as violently as before. Even in Sindh, where the provincial government was in the hands of the PPP, kidnapping, dacoity and political violence seemed on the increase, with the PPP unable to end the polarizations within society. Moreover, the military seemed reluctant to assist in the maintenance of law and order. Nor did the dismissal of the PPP government and dissolution of the National Assembly in August 1990 help to reduce violence in society, or restore the state's ability to protect its citizens.

6.4 Conclusion

Between 1977 and 1990 Pakistan experienced three structural types of militarism. First, from 1977 to 1985 the military was the ruling force, with a few civilians co-opted in; then, from 1985 to 1988 it was the decisive authority in co-partnership with civilians (with a brief reversion to being the main ruling force again after May 1988); third, from 1988 to 1990 the military maintained a decisive authority but under overall civilian supremacy.

Throughout the whole period the two main ideological strands were Islam and nationalism, but the bellicosity that might otherwise be expected to accompany the ideological dogmatism was tempered by a caution born of historical experience. As Mushahid Hussain points out (1990, p. 112), the Zia regime learnt a basic lesson from its two military predecessors: a military regime could not rule the people while at the same time fighting on the borders. Ayub's regime had weakened with the 1965 war, and the 1971 war signalled the end of the Yahya regime. In contrast, Zia kept the lines of communication open with New Delhi and Moscow, despite conflicts with both.

Finally, the focus of state politics under militarism followed the traditional pattern of high military expenditures regardless of the type of militarism that prevailed. Again, especially during direct military rule (1977–85), the military shaped the socio-economic goals and policy of the state and guided its internal security operations.

The linkage between militarism and militarization was also clearly evident in Pakistan during the 1977–90 period. However, the case of Pakistan also reveals that militarization of society had its own momentum, and its intensity did not run parallel to the degree of militarism. For instance, the election of a democratic civilian PPP government in Sindh in 1988 did little to quell the rising violence in Sindhi society. That the PPP government resorted to military power to deal with the near anarchic situation in Sindh shows how violence in civil society can compel the state to take recourse to violence. In such a situation, the militarization of civil society acts to sustain militarism within the state.

Notes

1. The development of Praetorianism has been discussed by a number of analysts, including Samuel Phillips Huntington (1957) and Samuel Edward Finer (1962).
2. In fact, the coup took place when an accord had already been reached between the PNA leadership and Bhutto, with the latter accepting most of the PNA demands. See Noman, 1988.
3. For example, Mir Aijaz Ali Talpur, *Herald*, January, 1988.
4. General Zia had only reluctantly agreed to the signing of the Geneva Accords on Afghanistan.

References

Alavi, H., 1989. 'Politics of Ethnicity in India and Pakistan', pp. 222–246 in H. Alavi & J. Harriss, eds. *Sociology of 'Developing Societies': South Asia*. London: Macmillan Education.

Askari, H.A., 1987. *The Military and Politics in Pakistan 1947–1986*. Lahore: Progressive Publishers.

Finer, S. E., 1962. *The Man on Horseback: the Role of the Military in Politics*. London: Pall Mall.

Gramsci, Antonio (Q. Hoare & G. Smith, eds), 1971. *Selections from the Prison Notebooks*. New York: International Publishers.

Herald, February 1986, January 1988, October 1988. Karachi.

Human Rights Commission of Pakistan, 1986. *The Sindh Report*. Lahore: HRCP.

Human Rights Commission of Pakistan, 1990. *Sindh Inquiry*. Lahore: HRCP.

Huntington, S. P., 1957. *The Soldier and the State*. Cambridge, MA: Belknap Press.

Hussain, A., 1989. 'The Crisis of State Power in Pakistan', pp. 199–236 in P. Wignaraja & A. Hussain, eds, *The Challenge in South Asia*. Karachi: Oxford University Press.

Hussain, M., 1990. *Pakistan's Politics: the Zia Years*. Lahore: Progressive Publishers.

Jalal, A., 1990. *The State of Martial Rule*. Cambridge: Cambridge University Press.

Noman, O., 1988. *Pakistan: Political and Economic History Since 1947*. London: Kegan Paul International Ltd.

Pinkney, R., 1990. *Right-Wing Military Government*. London: Pinter.

Thee, M., 1980. 'Militarism and Militarization in Contemporary International Relations', pp. 15–35 in A. Eide & M. Thee, eds, *Problems of Contemporary Militarism*. London: Croom Helm.

Wallensteen, P., J. Galtung & C. Portales, eds, 1985. *Global Militarization*. Boulder, CO: Westview Press.

7

Democratization in Bangladesh: the Mass Uprising of 1990 and Its Aftermath

MEGHNA GUHATHAKURTA

7.1 Introduction

Since its independence in 1971, Bangladesh has experienced successive coups and the consequent establishment of martial law regimes, with intermittent suspension of the Constitution. The history of democratization in Bangladesh has therefore taken the form of confrontational as well as constitutional politics – sometimes the one leading to the other, and sometimes the two existing in parallel. This has often led to volatile outbursts, and frequently culminated in turbulent movements, both of which have characterized the political scene in Bangladesh for the past two decades.

My concern here is to examine the dynamics of democratization in Bangladesh as it specifically relates to the nature and outcome of the December 1990 movement against the regime of President Ershad and the ensuing elections of 1991. I conclude with some comments on the problems and challenges facing democratization in Bangladesh.

7.2 A Brief Background

The birth of Bangladesh came about through a bloody confrontation between the Pakistani military junta and Bengali freedom fighters supported by India. In 1972, the Constitution of Bangladesh moulded the Bangladesh state into a parliamentary democracy, with the principles of nationalism, democracy, socialism and secularism set out as guidelines for running the newly independent state.[1] In economic terms a non-capitalist path of development was envisaged; whereas in political terms a nationalist bourgeois party (the Awami League) dominated by *petit bourgeois* elements created a situation which led to the plunder of state resources (Jahangir, 1986). Policies such as the creation of paramilitary forces (termed *Rakhhi Bahini*), alienated the military and forces of the radical Left alike. Both these forces turned against the government of Sheikh Mujib. In response, the government adopted repressive measures which resulted in the infamous Fourth Amendment to the Constitution, changing the government from a parliamentary democracy to a one-party presidential system with authoritarian powers vested in the head of state (Guhathakurta & Hossain, 1991). This sounded the death knell of the Mujib regime; in August 1975, the first elected government of Sheikh Mujib was toppled by a bloody military coup.

The coup hinted at intense factionalism and lack of discipline within the army. A series of counter-coups occurred in the first week of November 1975, from which Major General Zia-ur-Rahman emerged as the de facto ruler. He led the Martial Law Regime until 1977, when he floated his own party, the Bangladesh Nationalist Party (BNP), and held national elections through which he became the constitutional head of the Republic. Zia adopted various policy measures to stabilize and legitimate his rule. He tried to appease the army by increasing the defence budget from 7% to 20% of the national budget (Government . . . of Bangladesh, 1980, p. 363). He also followed an 'open arms' policy, inviting both right-wing Islamists and left-wing radical forces (who were united in their antagonism to the erstwhile centrist ruling party of the Awami League) into the rank and file of his supporters and ultimately giving them posts and positions in the government and party. But despite his populist pro-grammes, which managed to attract considerable foreign aid, he could not completely quell the dissatisfactions within the army. In May 1982, he was assassinated by rebels within the army.[2] His civilian government remained intact, but not for long. The rebels were soon hunted down and General Ershad, Commander-in-Chief of the Army, assumed control of the state.

7.2.1   The Ershad Regime

Martial Law was declared in March 1982. After a period in which the Constitution was suspended, General Ershad launched his own party, the Bangladesh Jatiyo Party, in the style of his predecessor, and called for general elections in 1986.

However, unlike Zia, Ershad chose to keep politicians out of his adminis-tration, appointing military and civilian bureaucrats to his cabinet of advisers. He also took measures to keep the military well satisfied by increasing their status and benefits – pay rises, subsidized rations, improved housing etc. Even as a Lieutenant General in 1981, Ershad had been known to make comments which indicated his preference for constitutionally incorporating a more active role for the Bangladesh military in the decision-making process.[3] He continued Zia's open arms policy, gathering together a following from mixed backgrounds in order to establish his political legitimacy. He also excelled as a 'political survivor', often withdrawing from vulnerable policy positions which caused undue consternation among the public, such as the reorganization of secondary education curricula, postponing the *upazila* (local bodies) decentralization programme etc. Most of all, his talents were concentrated on defusing any form of organized movement such as students or trade unions against his regime, by a clever combination of intimidation, concessions, persuasion, strategic divisiveness and selective resource alloca-tion (Khan, 1984). All the same, people took to the streets and eventually a formidable opposition grew in strength and size.

7.3 The December 1990 Mass Uprising

The mass uprising of December 1990 marked the culmination of the anti-Ershad movements which had started with the student protest against Ershad's education policy on 14 February 1983. In mid-March of that year, a 15-party alliance led by the Awami League and a 7-party alliance led by the BNP were established, forming the main channels for orchestrating demands for democratization. After experiencing three years of continuous protest, General Ershad announced a date for holding the National Parliamentary Election on 6 April 1985. Opposition parties still protested that their demands were not fully met. However at the insistence of the government, elections finally took place on 7 May 1986. For reasons still debated, the Awami League participated in the elections, together with seven of its alliance members. The BNP alliance and five parties of the original 15-party alliance continued their protest. However, soon after the Awami League realized that they were not making any headway from within Ershad's Parliament, they withdrew their support. As the movement strengthened, a liaison committee was formed which sought to bring together the three alliances on a common platform (the *Tin Oikyo Jote*) to demonstrate a unified show of resistance against the Ershad regime.

Another important factor which led to the mass uprising of December 1990 was the forging together of the *Chhatro Oikyo Jote*, a unified students' command which gave major leadership to the movement towards the end. This was a historical event because students, who have traditionally played an important role in the country's politics, had been prime targets of the 'divide and rule' policy of governments. Various tactics have been followed to create divisiveness among students, including the supply of arms, or playing upon party sensitivity, such as raising party-based slogans in the joint programme of the *Jote* (Roy, 1991). But despite every effort to divide them, the students managed to rise above their partisan politics and stick together, thereby providing an excellent example which other groups started following. Numerous cultural groups joined ranks under the banner of *Sangskritik Qikyo Jote*, while women's groups joined under the banner of *Oikyobodho Nari Samaj*. Professionals and, towards the end, even businessmen and government cadres took to the streets and declared non-cooperation with the Ershad regime.

Although questions have been raised as to whether the opposition movement had an urban middle-class bias, it is also true that it could draw on wide support from the urban poor and lower middle class. The reason for this might lie in the state of the economy during the Ershad regime. It was the urban poor and lower middle class who had to bear the brunt of the spiralling prices of essential goods – so much so that a *hartal* (general strike) was called by the opposition alliance on this issue on 16 October 1990. The strike was successful all over the country.

The following facts and figures will indicate the declining trend in the economy. Annual growth rates in the agricultural and industrial sectors in 1975–79 were 3.26% and 5.05% respectively: by 1981–89 they had fallen to 1.85% and

2.71% respectively. In 1980–81 domestic savings constituted 3.4% of domestic production, whereas in 1988–89, it declined to 1.1%. As a result foreign dependence increased. In 1981, per capita foreign aid was Taka 840, while by 1987 it had increased to T 2,720.[4]

Among the beneficiaries of the Ershad regime, the military occupied a prominent position. Here too some figures may be cited. In 1980–81, defence expenditures constituted 19.3% of budget allocations. In Ershad's time this figure became 25.2%. In 1987–88 it was slightly reduced, to 20.1%, on the insistence of aid donors. But rather than openly showing defence expenditures, many projects which actually benefited the military were simply moved to the development budget. If such allocations were included, then total defence expenditures would probably amount to 30% of the budget. Apart from this, the military also stood to benefit from the 'spoils distribution' policy of the regime, for example advantages in business contracts and government jobs. The preference given to the military over civilian personnel is clear from the Warrant of Preference issued by the Cabinet Secretary on 20 September 1986. Among other things, this warrant gives greater preference to the chiefs of the army, navy and air force than to Members of Parliament. Despite being a prime beneficiary of the Ershad regime, however, the military was to fail to give adequate support to Ershad when he needed it most.

Another feature which needs to be reviewed when discussing the dynamics of the movement is the role played by international donors. Bangladesh's dependence on foreign aid is an acknowledged fact which was not ignored by either Ershad or his opposition. Ershad demonstrated international support for his government by boasting of the volume of aid received. The opposition alliance in turn asked donors to stop aid to the Ershad government. It is also interesting to note that the movement of December 1990 differed from that of 1987 in that no anti-imperialist slogans were raised in December 1990. The slogan was for democracy; socialism now took a back seat. Furthermore, with the tide of the democratization process unleashed in the Soviet Union and East Europe, even the Army Commander-in-Chief thought that the government should give in to popular demands.[5]

Donors supported the opposition movement at a later stage. On 30 November 1990, the British and Japanese governments separately expressed their dissatisfaction over the state of affairs and declared that it might affect future aid disbursement (BBC Bangla News, 30 November 1990). Donor pressure could thus have been an influential factor in restraining the Ershad government from engaging in indiscriminate killing, harassment of civilians and other coercive practices.

From the above description we can generally conclude that the democratization process in Bangladesh was initiated and brought about by extra-constitutional means. The politics which led to the end of the Ershad regime were essentially a politics of rebellion. But although the movement was confrontational in character, its ultimate aim was not to overthrow the government by violent means, but rather to put pressure on the government to resign and make it transfer power to the people's representative. This intention was incorporated in an important document drafted by the three-part alliance liaison committee on 19 November 1990 (Uzzaman, 1992). The declaration emphasized that the immediate objective of the movement was to pressure Ershad and his government to resign and transfer power to a non-partisan and neutral representative of the people, to be nominated by the alliance. The mode in which this power transfer would be conducted was also specified. It was deemed that Ershad would appoint the nominated representative as Vice-President of the state, so that after the President's resignation, he could take over the Presidency. As head of a caretaker government, he would then have responsibility for holding free and fair elections within three months, to decide who should govern the country on behalf of the people. All this was quite innovative in the context of the political history of Bangladesh, and the Ershad regime was naturally reluctant to follow this process. But pressure from the movement was to prove too strong.

Aside from the immediate goal of initiating a peaceful transfer of power, the declaration incorporated some long-term objectives as well. These were: to set in motion a democratic process which would include the cultivation of values and ideals upholding the cause of the Liberation War of 1971; to set up a truly accountable government; and to free the people from any further threat of unconstitutional takeover through coups or assassinations. Further, power was to be transferred to a truly sovereign parliament, unlike the rubber-stamp parliament operating during Ershad's rule. This document appeared as the mandate of the people, and would in the following months constitute a central reference point in the people's continuous struggle for true representation.

Ershad's resignation left Bangladesh with a downtrodden economy and a political and administrative system ridden with corruption and malpractices. But the new caretaker government had to ignore these problems and get on with the task of organizing free and fair elections within 90 days. The problems were immense, and reforms had to be undertaken, especially in the administration, in order to ensure that elections could be truly free and fair. By and large, this goal was in fact achieved. I will now focus on some aspects of electoral politics which may have important implications for the democratization process in Bangladesh.

7.4 Parliamentary Elections 1991

As promised by acting president Justice Shahabuddin Ahmed (chosen by the alliance to head the interim government), nationwide parliamentary elections took place on 27 February 1991. Out of the 330 seats in Parliament, the BNP emerged with the single largest majority with 140 seats, the Awami League came second with 88 seats, followed respectively by Jatiyo Party (35 seats) and Jamaat-e-Islami (18 seats). Other seats were won by the various leftist parties, independents and well-known personalities from marginal parties (BAMNA, 1991).

A noticeable feature of the electoral politics which led to the polls was the virtual dismemberment of the unity struck by the opposition forces prior to Ershad's resignation, and the consequent development of partisanship. Violent clashes took place between the major contesting parties both prior to and after the elections (BAMNA, 1991).

Secondly, electioneering tactics concentrated on rhetorical issues rather than on policy matters. Most parties started campaigning long before their manifestos came out. Election campaigns were conducted on emotive lines, for example the Bangladeshi nationalism of the BNP vis-a-vis the Bengali nationalism of the Awami League, and the anti-Indian attitude of the BNP vis-a-vis the pro-Indian stance of the Awami League. Money too was a major factor in winning votes (BAMNA, 1991).

There were only a few reports of widescale rigging like stealing ballot boxes or 'capturing booths'. But some allegations of more subtle rigging and threats to the minority population were made, especially by Awami League candidates (*Daily Shongbad*, 4 March 1991). Voters' lists were reportedly drawn up with the help of Ershad's ward commissioners with the municipal elections in mind, and had deliberately omitted large sections of people who would possibly not have voted for Ershad. It was also alleged that the BNP had prior notice of this fact and hence could have used it to their advantage.

But despite such allegations, it is also true that in many cases, particularly in the rural areas, the outcome depended more on local power and influence. This would also explain why Ershad's Jatiyo Party could win so many seats (35), although the country had just experienced a movement condemning its government.

Party cadres played an important role in the campaign. The BNP managed to rally the support of college students, to whom issues like the ideals of the Liberation War belonged to history (and indeed a history with which they were not well acquainted, as it had been rewritten by successive regimes to suit their own purposes). Closer to their consciousness were the ideals of President Zia and the recent movement against Ershad. The Awami League, on the other hand, had longstanding support among the grass roots and a core of dedicated cadres, but these fell prey to factional rivalries, or over-confidence, or both.

7.5   The Post-Election Agenda

The election of 1991 did not solve any of the country's problems, although it gave Bangladesh a chance to re-democratize itself. But the pitfalls still exist.

The first post-election crisis to arise concerned the form of government. The Awami League had campaigned for a parliamentary form, while the BNP had advocated a presidential system with greater powers for the Parliament. According to a survey of 500 electoral candidates selected from 100 constituencies through random sampling, it was seen that about 63% of all candidates favoured a parliamentary form of government (*Gonotantrik*

Uddyog, 1991). Whilst even a month after the election, the top leaders of the ruling party (BNP) seemed to have been in favour of a presidential system, it reversed its opinion and openly advocated a parliamentary form of government (Speech of the Prime Minister, 1991). The reasons for this reversal may have been varied: pressure of public opinion, the disastrous cyclone of April 1991 that exposed the shaky foundations of the recently elected government, the risks involved in contesting a presidential election due to growing disenchantment among the populace. However, the real issue at hand was the question of state power. The Constitution of Bangladesh from the time of its Fourth Amendment in 1975 had conferred almost absolute power on the President, without any forms of checks and balance. No doubt the violent history and the ruptured political development of Bangladesh have caused the many amendments responsible for such a state of affairs (Guhathakurta & Hossain, 1991). During the mass uprising of 1990, the three coalition alliance declaration had emphasized that power be handed over to a sovereign parliament. For this the Constitution needed to be further amended.

The BNP and Awami League moved two separate constitutional amendment bills, on 2 and 4 July 1991 respectively. After several days of heated debate the two bills were referred to a Select Committee to reach a consensus on the issue. Finally on 6 July 1991, the Twelfth Amendment Bill was put to the vote twice. The results were 306 in favour and none against in the first round, and 307 in favour and none against in the second (Hakim, 1991–92). The Bill was thus uncontested and almost unanimously adopted. After 16 years of Presidential rule, Bangladesh returned to parliamentary democracy.

7.6   Conclusion

Determining the form of government is only the first step towards democratization. The process faces many pitfalls and challenges. And as we can see from the current political crisis facing Bangladesh, the path has certainly not been strewn with roses. While the government of Khaleda Zia nears the end of its first term, the main opposition parties have been boycotting the Parliament since March 1994, and have been involved in countrywide agitation against the ruling government. They have called for the government to resign and make way for a caretaker administration which would be responsible for the holding of free and fair elections. The ruling party is accused of being corrupt and of rigging votes. Such a situation has led to a state of political stalemate which is endangering the recently instituted process of democratization.

The failure of the democratic process to run smoothly in Bangladesh has been marked by the following characteristics: the failure to turn the Parliament into a truly sovereign body; the failure to make the executive truly accountable to the people, as well as ensuring the people's participation in state-run institutions; and the failure to implement the rule of law.

The smooth working of parliamentary politics has been impeded by several factors: an order of precedence, both formal and informal, carried over from the autocratic rule of Ershad, which gave the military more importance than Members of Parliament; the intrusion of personalized politics in the affairs of the state in the absence of a consensus between ruling and opposition parties. Many have been the times when parties have staged a walk-out or the Parliament has reached a deadlock because one member insulted another with a personal remark and then refused to apologize. All this has contributed to making the Parliament an ineffective body in which to resolve the affairs of the state, which in turn has left many of these unresolved issues to be settled by 'street politics' – through demonstrations of strength such as calling strikes, mass meetings etc.

It is not perhaps odd for the executive to demonstrate its strength in any democratic state by making political appointments. What is unusual, however, is when the government crosses accepted boundaries and interferes in the smooth running of the bureaucracy, and even non-governmental institutions in education, research and the financial sectors – which is what the ruling party in Bangladesh has been accused of doing. The ultimate expression of this is, of course, to be found in the massive rigging of votes in several by-elections, the most publicized among them being in the constituency of Magura in the southern district of Faridpur (*Morning Sun*, 10 April 1994).

Although, in terms of the freedom of the press, the print media have shown some signs of liberalization, radio and television remain under governmental control. Also, indirect pressure is often exerted on the print media through government control of advertisements. The refusal of the government to acknowledge as legitimate many of the demands arising from the society (for example, the demand to hold trials for the war criminals of 1971, some of whom are Members of Parliament and political leaders of the fundamentalist party Jamaat-e-Islami), has aggravated tensions in the system, often leading to a virtual breakdown of law and order.

The most incompetent performance of the present government has been in its failure to establish a rule of law (Hossain, 1994). The situation has hardly differed from the preceding autocratic rule of Ershad. Terrorism and extortion practised by armed gangs of *mastans* (thugs) has been one of the obnoxious features of the old order which has continued into the present – possibly because not a single political party has been averse to using these forces to pursue their own interests.

With the political stalemate in 1994 being resolved through the help of external mediation, an emisary from the Commonwealth Secretariat, the concerned parties have so far shown a willingness to participate in dialogue – but to what extent they are willing to compromise is yet to be determined. From a long-term perspective, the future of constitutional politics rests on the quick and effective resolution of the socio-economic problems of the country, or else the national consensus so necessary for the functioning of parliamentary democracy may soon dwindle. In that case, there is every possibility that the politics of confrontation will again push their way to the forefront in Bangladesh.

Notes

1. For an account of the background history of the birth of Bangladesh and the initial years after independence, see T. Maniruzzaman (1980); R. Jahan (1980); and Zillur R. Khan (1984).

2. There are several versions of Zia's death. Some say it was a personally motivated murder, but others interpret it as part of an overall plan to exterminate all freedom fighters in the army as opposed to the repatriates from Pakistan. See Zillur R. Khan, 1984, pp. 225–230.

3. Statement released to the Bangladesh press on 28 November 1981.

4. Quoted from an unpublished article by Binayak Sen in Muntassir Mamun (1991, pp. 39–43). Also see Rehman Sobhan (1991).

5. Television interview with General Nooruddin, Army Chief of Staff, 10 December 1990.

References

BAMNA (Bangladesh Mukts Nirbachon Anddan), 1991. *Election Watch Report*. Dhaka: BAMNA.

Gonotantrik Uddyog (Democratic Initiative), 1991. *A Survey of Electoral Candidates of the National Parliamentary Elections*. Dhaka: Gonotantrik Uddyog.

Government of the People's Republic of Bangladesh, 1980. *Statistical Year Book of Bangladesh*. Dhaka: Bureau of Statistics.

Guhathakurta, M. & S. Hossain, 1991. 'Bhagyer Shongbidhan Na Shongbidhaner Bhagya' [An Account of Constitutional Amendments in Bangladesh], *Rupantor*, June.

Hakim, Muhammad A., 1991–92. 'The Fall of the Ershad Regime and its Aftermath', *Regional Studies*, vol. 10, no. 1. Winter.

Hossain, K., 1994, 'Transition of Democracy', in *South Asia: Vision and Perspectives*. Lahore: South Asian Regional Dialogue.

Jahan, R., 1980. *Bangladesh Politics: Problems and Issues*. Dhaka: UPL.

Jahangir, B.K., 1986. *Problematics of Nationalism in Bangladesh*. Dhaka: Centre for Social Studies (CSS).

Khan, Zillur R., 1984. *Martial Law to Martial Law: Leadership in Crisis*. Dhaka: UPL.

Mamun, Muntassir, 1991. *Shob Shombhober Deshey*. Dhaka: Pallab Publishers.

Maniruzzaman, T., 1980. *The Bangladesh Revolution and Its Aftermath*. Dhaka: UPL.

Roy, Swadesh, ed., 1991. *Gono Abuthyan '90*. Dhaka: Pallab Publishers.

Sobhan, Rehman, ed., 1991. *The Decade of Stagnation: the State of the Bangladesh Economy in the 1980s*. Dhaka: UPL.

Speech of the Prime Minister (Khaleda Zia) to the Nation, 1 July 1991, Dhaka.

Uzzaman, Hasan, 1992. *Amader Bhobishot O Koronio: Nobboyer Joutho Ghoshanar Alokey*. Dhaka: Pallab Publishers.

8

Militarization, Violent State, Violent Society: Sri Lanka

JAYADEVA UYANGODA

8.1 Introduction

In Sri Lanka, and in South Asia in general, recent decades have witnessed a fundamental shift in the post-colonial political order, with violence re-emerging as the main characterizing factor in state–society relations. To say that the state has increasingly become violent is both a truism and an understatement. Rather, violence no longer constitutes a hidden dimension of state power, as it used to under the brief liberal-democratic interlude of two or three decades after Independence. Similarly, the way in which subordinate ethnic groups and social classes relate to each other and negotiate with the state has come to be characterized by a recurrent propensity to use collective violence as a means of articulating their demands.[1]

Social groups and political forces do not appear to see electoral and parliamentary competition as an effective, viable means of political mobilization. The tendency to use violence for political objectives now exists alongside conventional adherence to the rules of parliamentary competition and electoral bargaining. State-centric and counter-state violence represent a qualitative shift in the dominant mode of political bargaining in South Asian societies today.[2] Perhaps, violence has become the governing idiom as well as the praxis in the post-modern political condition in South Asia.

On the other hand, it would be a historical fallacy to assume that political violence is a totally new phenomenon in South Asian societies. In pre-colonial as well as colonial periods, militarized violence, with abundant material and ideological resources, was certainly used by the state in order to ensure the subjugation and domestication of the masses. Considering the frequency of warfare that characterized the pre-colonial political conflicts in South Asian social formations, one cannot fail to recognize the intensely militarized nature of state–society relations as these that have evolved through pre-mediaeval, mediaeval and colonial polities.[3] The colonial conquests, and their aftermath as well, produced a militarized and violent political order in which the state was re-constituted as a highly efficient and rationally organized force for social subjugation. Viewed in a broad historical context, the contemporary violent state is a post-colonial reconstitution of a polity that had exploited the utility of violence through all phases of its transformation.

What is so contemporary about the presence of violence in the general political practice in Sri Lanka? Indeed, this question is not unique to Sri Lanka: it is equally relevant to many post-colonial societies where optimistic visions of an eventual transition to liberal democratic polities were shared by both the ruling and intellectual elites at the time of political independence.

This political optimism has mostly been proved untenable, in various ways. Even in societies where parliamentary democracy has not yet been overthrown by authoritarian forces of either the Right or the Left, the continuity of formal and widely supported democratic institutions largely conceals the role played by non-democratic understructures of political relations at both state and counter-state levels.

In India and Sri Lanka, for instance, institutional arrangements of the democratic state have survived many political conflicts and tensions; yet they no longer rest on the same social consensus or social contract that was negotiated (and at times re-negotiated) in the aftermath of Independence. Now are those democratic institutions – formal, constitutionally grounded and associated with parliamentary democracy – any longer the sole or the most efficient and effective networks to constitute state–society linkages? They have already been paralleled, supplemented and even subsumed by a host of extra-constitutional and extra-legal organs of power – death squads and vigilante groups – along with increasing deployment of the military to fight counter-insurgency wars.

The formal institutions of state power exist side by side with recently emerged 'un-formalized' agencies of state violence through which questions of the legality, constitutionality and accountability of a variety of state practices can be circumvented. For example, the continuity of the formal institutions of the judiciary, supporting the idea of the rule of law, merely masks the very real phenomenon of politically necessitated extrajudicial killings by official and unofficial death squads. The point then is that the repressive apparatus of the state can no longer be treated as secondary to formal organs of the state such as the legislature, the executive and the judiciary: they *are* substantially the state.

Moreover, rebellious social and ethnic groups have been rejecting the relevance of formal practices of parliamentary democracy in their political programmes. Take, for instance, the rejection of parliamentary elections by militant Akali groups in Punjab, India or by the Liberation Tigers of Tamil Eelam (LTTE) and the Janatha Vimukthi Peramuna (JVP) in Sri Lanka. The fact that these and many other political movements have chosen the path of armed struggle to press for their demands and express their active and violent opposition to existing institutional organizations of political bargaining is symptomatic of a profound crisis in the formal institutions of the state.

This crisis and the manner of its unfolding can be outlined in somewhat abstract terms: there is a decisive breakdown of the post-colonial political consensus as demonstrated by the moving away of a significant section of society – ethnic minorities and intermediate social strata – from the state. In other words, a wide rupture has occurred between the state and society, creating a situation in which the state finds it exceedingly difficult to command passive social loyalty and obedience. Healing of this rupture requires a new social consensus capable of (1) making the state responsive to the demands of rebellious social and ethnic groups, (2) bringing back to the state those groups and classes that have distanced themselves from the state and (3) changing the class and ethnic character of the state.

These three requirements imply the need for a comprehensive reform project, signifying a radical revision of the post-colonial social contract. The vital issue today concerns the necessity of a new social contract. Looking at the political conflicts of the past decade or so, we may notice that the broad framework of a new social contract is actually being negotiated by means of violence. For example, the current phase of the war between the Sri Lankan state and the Liberation Tigers of Tamil Eelam has been primarily concerned with how to devolve and share state power in a restructured polity.

In this sense violence is also a means by which the contending parties are engaged in the negotiating process. The outcome is not yet clear, because the negotiation is still taking place in essentialist and mutually exclusivist discourses. The paradox of violence manifested in this particular context is that ethnic essentialism breeds violence, and violence, in turn, reinforces essentialism and exclusivity.

8.2 Phenomenology of Political Violence

What we have witnessed in Sri Lanka over the past few years is also a process of violence being normalized in individual as well as social relations. A wide range of power-brokers have begun to impose their competing wills on society, and violent practices have became normal and regular.

The militarization of social and individual relations in Sri Lanka has also been accompanied by militarization of political conflicts. There are two main ones which have assumed an exclusively militaristic character. The first involves the *ethnic question*.

The pattern of the conflict in the 1980s was that the state, Sinhalese nationalist forces and Tamil nationalist groups came to regard armed confrontation as the only plausible mechanism to resolve ethnic antagonisms. Even in the early 1990s, irrespective of occasional utterances to the contrary by parties to the conflict, the advocacy and practice of a military solution remained so pervasive that there was very little social space for a non-militaristic alternative to the ethnic war. In short, political debate too has become militarized and the categories of political language have tended to signify uncompromisingly militaristic assumptions.

The second political conflict, which also contained essentially the same militaristic dimensions, was that between the state and the Janatha Vimukthi Peramuna (JVP) in the late 1980s. Essentially a nationalist protest rebellion launched by youth groups in Sinhalese society, the JVP-led political campaign and the way the state responded to it brought to the fore the fact that once a conflict is militarized, de-militarizing it is enormously difficult. Indeed, this conflict was ultimately 'resolved' in late 1989 by near total physical annihilation of the JVP.

This process, in the meantime, exposed some characteristics of political violence in a militarized context. It demonstrated, first of all, that the outcome of the conflict would depend largely on the capacity and willingness of the protagonists to be ruthlessly brutal and violent, not only against each other, but also against a society reduced to the status of a terrorized onlooker. It also showed that the capacity of the state to use violence against its domestic adversaries had virtually no bounds – material, political or moral.

8.3   Social Acceptance of Political Violence

Any observer of Sri Lanka's political conflicts is certain to notice the ease with which society has accepted political violence as a legitimate mode of political behaviour, whether by the state or anti-state forces. This mass legitimation of political violence is symptomatic of the present incapacity of Sri Lankan society to produce indigenous arguments for non-violence, despite the professedly non-violent moral codes of its dominant religion, Buddhism. Ironically, contemporary Sinhalese Buddhism has moved towards openly advocating state violence of the ethnic kind; it has openly campaigned for intensifying the war effort of the state against Tamil separatist movements.[4]

Living through the extremely violent political conflict of 1988–89 in the Southern part of Sri Lanka, I myself was particularly astonished to see how people even *admired* politically motivated brutalities practised by young JVP rebels and by members of the state forces. Still more perplexing was the advocacy by some leading Buddhist monks of a military solution, as opposed to any form of political settlement of the Tamil ethnic problem.

This makes it extremely difficult for the student of Sri Lankan politics to separate the violent state from a society which arguably possesses a powerful ideological understructure of political violence.[5] To put it slightly differently, violence has become a norm in the contemporary political culture, and it is a phenomenon normalized in all forms of political practices of the state as well as of counter-state formations. One disturbing aspect of the social acceptance of political violence is that it prepares social space for long-term processes of militarization at various levels.

How has the mass acceptance of political violence, practised by the state and anti-state forces alike, evolved in the recent political history of Sri Lanka? Until the late 1970s, Sri Lanka experienced relative social peace, except on two occasions: the anti-Tamil riots of 1958 and the youth insurgency of April 1971. The ethnic riots of 1958 were fairly widespread; for the first time in independent Sri Lanka, both state violence and mob violence were deployed against Tamil civilians.[6] The insurgency of 1971 was a different manifestation of political violence: an attempt made by a political movement of Sinhalese youth to capture state power through an armed rebellion.[7] On that occasion, political violence was essentially directed against the state to secure the goal of political power. The insurrection was swiftly put down by the state by means of violence which was at the disposal of the state.[8]

Popular responses to these two episodes contained what could be observed subsequently in the 1980s in the form of a mass phenomenon, namely the justification and tolerance of political violence from a variety of perspectives. Three of these perspectives can be delineated as follows: (1) state as well as mob violence directed against ethnic minorities, particularly Tamils, as just and legitimate; (2) organized violence against the state as an act of heroism; and (3) the use of excessive political violence by the state, however undesirable, as yet something with which society should somehow or other come to terms.

8.3.1 Examples of Violence

Let us now see how these perspectives have been present in some key political conflicts during the past three decades.

The Anti-Tamil Riots of 1958   The anti-Tamil riots of 1958 were remarkable for the manner in which Sinhalese society accepted them as triumphant violence. Except for some sporadic stories about some Sinhalese who sheltered Tamils from Sinhalese mobs, there is hardly any popular memory or sentiment in Sinhalese society that could be construed as being at least sympathetic to Tamil victims of ethnic violence. On the contrary, there are many anecdotes celebrating events, real or imaginary, of killing, torturing and burning of Tamil civilians in predominantly Sinhalese areas.

In the Sinhalese political consciousness, the violence of 1958 is regarded with pride. Violence was the means by which Tamils who had exceeded the behavioural limits of an ethnic minority could be put back in their 'proper' subordinate place. In a society where ethnic relations are hierarchically ordered, ethnic violence then was also a structural mechanism of re-establishing domination and submission. The events in the post-1983 years consistently affirmed this particular assumption inherent in the Sinhalese world-view.

The Insurrection of 1971   The insurrection of 1971, meanwhile, kindled sentiments of a specific form of romanticized anti-state violence. Being the first event of anti-state mobilization aimed at taking over state power, the insurrection was viewed by the general populace as a symbol of youthful idealism, courage and selfless commitment to a cause. Even today, more than two decades after the insurrection, the public memory of it is one of deep sympathy and identification with the youth who are seen as having rebelled against what were popularly called *samaja asadaranakam* (social injustices). It is perhaps paradoxical that the working-class uprising of 1953, a relatively peaceful *hartal*, has not evoked such powerful and romanticized public memories.

It seems to me that political mass consciousness of politics generally possesses a capacity to idealize, and not necessarily to disengage itself from, anti-state violence. In order to critique anti-state political violence, any society would need powerful ideological and moral codes of political non-violence. If the critique of counter-state violence comes from the state, it will have no moral argument against violence in politics, because the state's morality of non-violence is a mere re-affirmation of its monopolization of violence and terror as means to achieve political goals. This, ironically, is what happened in the left-wing political debate in the aftermath of the insurgency of 1971.

The mainstream Left, which even then enjoyed a leadership position in the dominant political debate, criticized the insurrectionary violence of the JVP from the viewpoint of the state. Being partners in a 'socialist' coalition government with the Sri Lanka Freedom Party (SLFP), the mainstream Left argued against the JVP from the standpoint that the JVP's political violence constituted counter-revolutionary terrorism, and not revolutionary violence à la Bolshevism.[9] The question of violence was then narrowed down in the left-wing debate to one of how, when and against whom violence should be 'correctly' exercised, and not to one of questioning its politico-moral bases.

The aftermath of the 1971 insurrection also witnessed how even a confrontation of limited violence between the state and anti-state forces could ultimately strengthen and expand the institutional bases of state violence. It was indeed during this confrontation that the state argument for political violence expanded to include the moralistic argument that the state had a legitimate right to exercise violence to protect the people, and that that right should be protected from public scrutiny.

Until then, there had perhaps been some reluctance on the part of regimes to resort to killings and arrests on a massive scale – not necessarily because a need had not arisen in the past, but because the state did not possess the ideological arguments to unleash its resources and institutions of violence. In the post-1971 years, however, any minor political provocation was certain to evoke the violent responses of the state. The insurrection of 1971 thus marked the coming of age of the Sri Lankan state.

Political Violence in the 1980s   In the 1980s, political violence took a qualitatively new turn. The dominant political debate was essentially conducted by resorting to, and generating, mass legitimacy for means of unrestrained violence.

The ethnic war was the first civil war to flare up in Sri Lanka in the 1980s. In both Sinhalese and Tamil societies violence became the essence of privileged political discourse, although based on different politico-moral arguments. In Sinhalese society, the task of defending the sanctity of the state (the latter being identified with the Sinhalese political hegemony) was transferred away from the traditional institutions of political bargaining and mediation: the agencies of war were brought in to conduct the ethnic debate.

A parallel process unfolded in Tamil society. The central argument there was that an unresponsive and racist state had to be dismembered by means of war in order to satisfy minority ethnic aspirations.

The militarization of the ethnic conflict in the 1980s had several dimensions. First, Sinhalese–Tamil ethnic relations became militaristic, a development which was to have far-reaching implications for Sri Lankan politics. Even today, the fundamental question concerning peace in Sri Lanka revolves around a single question: how to de-militarize the ethnic issue. The failure of both the Indo-Lanka Accord and subsequent peace negotiations between the state and the LTTE indicated that the political means remained subservient to military objectives.[10] This twin failure of peace attempts has paradoxically resulted in re-militarization of the ethnic question.

Militarization thus suggests more than the strengthening of militaristic institutions directly engaged in political conflicts. It also means the social acceptance of militarism as *the* legitimate and correct form of political practice in cases of crisis.

Let me illustrate this point by referring to changes in the mass political consciousness concerning ethnic violence, as demonstrated in the 1980s. In Sinhalese society, the view shared by the majority of political forces, members of moral (religious) communities, the intelligentsia and the general populace was that only a military victory over Tamil rebels could ultimately secure the territorial integrity of Sri Lanka. The fulfilment of the Sinhalese-Buddhist nationalist project was thus entrusted to the armed forces. New practices of Buddhist religious rituals – *bodhi pooja* – were invented by zealous politicians and their priestly cohorts to involve the masses in a new politico-religious culture of war.[11] The central message, readily accepted by nationalistically mobilized masses, was that this ethnic war was a 'Holy War' of epic proportions and with historical antecedents in the chronicles of Sri Lanka. Memories recorded in mytho-history were readily evoked to prove that the war of the state was just and imperative: a historical necessity.

It is this combination of state power, religion, mythology and popular ethnic prejudice which laid the firm ideological and popular foundation for the militarization of the state in the 1980s. The point is that this militarization of the state began not in conflict with the political aspirations of the Sinhalese masses, but indeed with their blessing. The militaristic capacity of the state – along with the passing of repressive legislation, the will to conduct a prolonged war and the increasing violations of human rights – increased and did not occur in social or cultural isolation. Heightened Sinhalese nationalism, with its ideological arguments for political supremacy, acted as the mediatory link between the militarized state and Sinhalese society.

The Second JVP Insurrection   The second insurrection of the JVP in post-1987 years brought to the surface some of the sharp contradictions in the above outlined process of militarization. First of all, the JVP's plan for achieving power was executed through a strategy of protracted armed rebellion. That was based on the Sinhalese idea of nationalism. The central political argument of the JVP against the state was that the latter did not possess the capacity and will to take the ethnic war to its logical end – a military victory over Tamil rebels. This argument was shared by almost all the Sinhalese opposition parties, with the exception of the Left. And it found a supportive constituency among the Sinhalese masses as well.

The JVP's point was that the state was not adequately militaristic with regard to the ethnic question. In an interesting twist of events, the extreme Sinhalese nationalist forces came to the above conclusion at a time when a section of the UNP regime was showing its willingness, agreeing to a system of devolution, to try out a non-militaristic strategy to resolve the ethnic question.[12]

The paradox here was that a regime which before had pursued a military solution to the ethnic question came under vigorous attack by its own ideological constituency when the regime began to move away, even to a limited degree, from a military solution. The JVP's nationalistic rebellion and the massive public sympathy it received in the initial stages of its insurrection clearly demonstrated that militaristic nationalism is a trap in which the Sri Lankan state today is caught. However, the decimation of the JVP as a political movement does not necessarily mean that nationalistic militarism too has disappeared from the political culture of Sinhalese society. It is cherished and propagated by influential groups of Sinhalese intellectuals.[13]

Both Tamil and Sinhalese societies, the former more than the latter, experienced a generalized state of violence and terror during most of the past decade. Tamil society continues to undergo the trauma of violence in war. The contemporary political experience in both societies points to sheer incapacity of all actors in the political conflict to bring an end to political violence. What puzzles the student of modern Sri Lankan politics is the monumental failure of all political movements to come to a sober assessment of the viability of violence as a means of achieving political and social change.

We may take the example of the JVP in the South to illustrate some aspects of this failure. In the post-1987 years, the JVP used violence on such a widespread scale that there were occasions in 1988 and 1989 when the state appeared on the verge of collapse. Indeed, the JVP managed with tremendous success to break the monopoly of violence formerly enjoyed by the state. Initially, this destabilization of the state evoked public celebration. Romanticization of the JVP's anti-state violence seemed a rather gratifying experience in the daily life of many who had opposed the United National Party regime which had been in power uninterruptedly since 1977. What I was told one day in October 1988 by a small shopkeeper in Colombo illustrates that. He said, 'This government listens only to the language of the gun.'

In the end the destabilized state did not collapse. Instead, it regained its militaristic and violent vitality with enormous force and vigour in mid-1989, demonstrating that in the exercise of violence the state could be far more brutally innovative than the rebels.

8.4 Violent State

Very little has been documented or written concerning recent changes in Sri Lanka, especially about its expansion of programmes and structures for repression. A decade of civil war and social rebellions has indeed made the state a militaristically interventionist force. This goes beyond the acquisition by the state of certain repressive capabilities through such legislative means as Emergency Regulations and the Prevention of Terrorism Act. It concerns the emergence of new informal methods of repression and militaristic intervention: secret methods that exist side by side with formal and legalistic ones.

The political significance of underground or secret structures of the state, specifically to facilitate undercover operations of abduction and execution of rebels and political opponents, can hardly be exaggerated. Exact information about these underground groups is hard to come by, although it is publicly known that death squads became particularly active in combating the JVP rebellion in the South in mid-1988.[14]

The names of these groups usually appeared on posters that were publicly displayed. Some of them were the 'Black Cats', *Ukussa* (eagle), or PRAA (People's Revolutionary Red Army). Official denial of their existence notwithstanding – to be underground, these state organs were supposed to be unidentified, unnamed and unacknowledged! – highly secretive military operations carried out by these death squads and other paramilitary groups intended to compete with the reign of terror were unleashed by the JVP.[15]

The *modus operandi* of these secret death squads, according to the available knowledge about them, was geared to achieving two objectives: to apprehend and kill suspected members of the JVP, and to instil as much fear as possible in the public mind. As many accounts of their activities indicate, these groups used to travel in vehicles without number plates, often in civilian clothes. There were also reports that they wore black clothes and black masks. Although unidentifiability was initially important in their operations, these groups later made it a point to make the public aware of their activities. Almost daily, burnt bodies of victims, mostly in groups and mutilated beyond recognition, were left in public places, along with notices acknowledging responsibility for the killings.

A brief recapitulation of events in 1989, the year in which these underground organizations were vigorously activated, would be useful for us to understand the conjuncture of the qualitative rise in state violence. Presidential elections had been held in December 1988 and the new President, Ranasinghe Premadasa, assumed office in January 1989. A month later came Parliamentary polls. The United National Party, in power since 1977 and now under the leadership of Mr Premadasa, secured a majority in the legislature. The JVP, which by this time had mobilized extensive public support and had already launched its armed struggle for power, boycotted and disrupted these elections.

As a result of the JVP's violence and the military activities of the Indian Army in the North-East provinces, voter turnout was quite low. This created a rather delicate legitimacy crisis for the state because the state could not marshall adequate public participation at the legitimacy-seeking exercise of elections at a time when its authority was being frontally attacked by the JVP. The JVP's strategy was to undermine the authority of the state by imposing its own authority on society by means of violence, terror and armed intimidation, acting as a parallel or quasi-state.

In June 1989, the JVP called upon the armed forces to defect from the state, obviously making plans for a decisive, if not final, assault. And indeed, a quasi dual-power situation had already been developing, its main feature being people's acceptance of the JVP as the emerging government. Central in this situation of disequilibrium was the ability of the JVP rebels to break the state's monopoly over violence and militaristic power by means of its own generalized practice of violence and terror.

Quite interestingly, people admired the JVP not because it implemented any economic or social reform programmes in areas where it held sway, but because the rebels displayed their militaristic prowess by the indiscriminate killing of those branded as enemies or traitors.

By June–July 1989 it was becoming quite clear that the stability of the state could no longer be secured by conventional methods of military action alone. The government was perhaps not certain about the loyalty of its own armed forces. A weakened state then began trying to re-establish its authority by creating and deploying new organizations for violence. It is against this background that death squads and paramilitary groups heightened their operations in July 1989; within four months they succeeded in hunting down the entire leadership of the JVP and large numbers of its members.

It is perhaps useful to compare the nature of the counter-insurgency war in 1971 with that of 1987–89. The JVP insurrection of April 1971 was a rather brief affair, extending for not more than three weeks. In contrast, the second rebellion was a protracted one, covering a period of over two and half years. In 1971, the JVP had no mass basis or mass mobilizational strategy at all, which was in sharp contrast with its post-1987 campaign. And indeed, it was the extensive mass base of the second JVP insurrection which to a large degree explains the phenomenal increase in violence deployed by the state in its counter-insurgency campaign in 1988–89. In 1971, the state managed to suppress the insurgency of essentially inexperienced young rebels with only the help of conventional forces. There were no reports of secret death squads functioning in April 1971, nor did the JVP resort to large-scale terror and violence as it did during its second rebellion.

8.5 Democracy and De-militarization: Renewal of the Democratic Debate

Sri Lanka's political debate on democratization has been moving along a path of transformation during the past few years, with some significant and positive results. The space within which both the terms and the content of the

democracy debate could assume a wider political significance was produced
by a series of events and processes. Notable among them are the following.

8.5.1   The Ending of the JVP Insurrection in the South

It is a gigantic irony that the extremely violent manner in which the JVP's
insurrection spread and later was put down by the state has served to create
social space for the revival of peacetime politics in Sri Lanka. Public resentment over the political violence of the late 1980s was felt at at least two levels:
first, the romanticization of counter-state violence lost its place in society,
making it difficult for insurrectionary politics to make a comeback for some
time; secondly, the public critique of the state too began primarily to revolve
round the rejection of state terror and violence.

8.5.2   Breaking up the Monolithic Unity of the UNP Regime

An event with far-reaching political consequences was the crisis that developed within the UNP regime in 1991, resulting in the weakening of the
regime's oppressive grip over the entire society. In August, a powerful section
of the ruling UNP and the Opposition parties in Parliament made an attempt
to remove President Ranasinghe Premadasa from office by means of an
impeachment motion. Although this attempt failed and Premadasa managed
to retain power, the challenge effectively put an end to the myth of the invul-
nerability of his regime. Consequently, the fear that had earlier governed
relations between it and society largely disappeared, and the Opposition
could attract growing public participation and support, thereby creating a
new opening for the democracy debate.

8.5.3   Revival of Democratic Civil Society

One of the most notable and encouraging features of Sri Lanka's politics, at
least since 1991, has been the invigoration of the autonomous democratic
movements and efforts, along with a considerable rise in the democratic consciousness of the masses. Perhaps the demise of the JVP, which sought to
provide a violent political alternative to the regime and the weakening of the
regime, in a situation of internal crisis removed the obstacles to politics within
civil society. Demands and agitation for the widening of democratic space in
this period centred on media freedom, improvement in the human rights situation and democratic constitutional reforms. The regeneration of the
democratic debate in turn enabled the parliamentary opposition, which had
remained unassertive during the preceding period, to make a comeback in
electoral politics in 1993.

8.5.4   Weakening of the Argument for a Military Solution to
Ethnic Conflict

If in the 1980s political debate on the ethnic question was primarily hegemo-
nized by the militaristic schools of Sinhalese and Tamil nationalisms, the

first few years of the 1990s saw gradual social acceptance of the idea of a negotiated and peaceful alternative to the war. The main reason for the sustenance of the 'military option only' argument had been the widespread belief in both societies that the ethnic war was winnable and therefore should be fought until a victorious outcome was attained. Although the main parties to the war – the state and the Liberation Tigers of Tamil Eelam – continued with the military campaign to achieve their respective objectives, a decade of protracted war had diminished the people's enthusiasm and support for the war effort. Perhaps a war-weary people could no longer respond to the fighting in ethno-romantic terms.

8.6   Conclusion

Sri Lanka's experience in recent political changes indicates at least two vital lessons concerning the democratic process, in a context of militarized political conflicts. First, internal conflicts, once they have entered the path of armed confrontation between the state and counter-state forces, acquire a particular reproductive dynamic, restricting the social space for the democratic argument. Further, the first period of such militarized conflicts may generate positive social hopes and responses for militaristic practices. Secondly, militarized conflicts, in the course of their destructive passage to maturity, may also create another dynamic: namely, the social isolation of the arguments for the continuity of violence. Therefore, the best guarantee for the success of democratic transition – with demilitarization of political conflicts, demilitarization of the state, rebuilding of democratic institutions and practices etc. – is the struggle for autonomous and democratic civil society politics.

Notes

1. Ethnic conflicts in Sri Lanka and ethnic, religious and caste conflicts in India are the most visible instances of violent intergroup relations.

2. I use the term 'counter-state' to denote those armed political movements, nationalist as well as 'left-wing,' whose projects are meant to seize political power by means of violence.

3. Historical records in India and Sri Lanka are replete with accounts of constant warfare among rulers from the early days of state formation. A general conclusion concerning the state formation in pre-colonial South Asia – it may apply to many other pre-capitalist societies as well – is that the state and politics had been transformed by a process of militarization. Even the great eras of peace as described in traditional historical records – for example, the reigns of Emperor Asoka in India and King Dutugemunu in Sri Lanka – were made possible after enormous state violence resulting in mass carnage.

4. For an account of Buddhist ideological campaign for intensifying the current war effort of the state, in opposition to a negotiated political settlement, see 'Sinhala Nationalist Reaction . . . in *Thondaman Proposals*', *Pravada*, February 1992, pp. 5–8.

5. Going against the grain of conventional wisdom concerning non-violence in Sinhalese-Buddhist culture, I hold the view that Sinhalese Buddhism has made no significant contribution to the evolution of a non-violent social ideology. On the contrary, the Sinhalese Buddhist historiographical tradition and ideology inherent in it supports ethnic political violence. The

dominant Sinhalese-Buddhist ideology also presupposes an authoritarian state, and not democracy and pluralism.

6. For an impassioned account of ethnic violence in 1958, see Vittachi (1958).

7. Among useful, though not necessarily insightful, works on the JVP insurgency are Alles (1976); Chandraprema (1990); Gunaratne (1990).

8. The actual number of JVP members and sympathizers who were killed in 1971 is not yet known. Estimates, mostly based on impressionistic calculations, vary from 5,000 to 20,000. A little over 18,000 JVP rebels were captured and imprisoned.

9. Members of the coalition government with the Sri Lanka Freedom Party were the Trotskyite Lanka Sama Samaja Party (LSSP) and the pro-Moscow Communist Party of Sri Lanka (CPSL).

10. The Indo-Lanka Accord of July 1987, signed by Prime Minister Rajiv Gandhi of India and President Junius Jayewardene of Sri Lanka, intended to end Sri Lanka's military conflict by creating a system of devolution in the form of provincial councils. The Accord also enabled the Indian Army to intervene in Sri Lanka's North-East in order to disarm Tamil militants. However, the Accord resulted in creating new political and military difficulties for both India and Sri Lanka, because the LTTE, the dominant Tamil armed group, refused to go along with the terms of the Indo-Lanka agreement. While India got itself embroiled in an unwinnable war with the LTTE for two years, in the South, Sinhalese nationalist opposition spread dramatically creating greater political instability. For a detailed account of the Accord and its aftermath, see Shelton Kodikara (1989).

11. *Bodhi Poojas* among Sinhalese Buddhists have traditionally been a form of congregational worship of the Bodhi tree. However, this practice became politicized, or rather politicians and the state appropriated it, in the mid-1980s as a means of Sinhalese mass mobilization in support of the war effort. For a fairly comprehensive discussion on the origin and spread of this state religious practice in the 1980s, see Kapferer (1988).

12. The Indo-Lanka Accord of 1987 was the expression of this change of attitude among a section of the UNP regime under Junius Jayewardene. For a discussion of this particular point, see Uyangoda (1989).

13. For a critical account of these intellectual groups, see Uyangoda (1992).

14. Some accounts of these paramilitary groups are provided in Chandraprema (1990) and Rohan Gunaratne (1990).

15. A quite useful, though limited in scope, public debate has emerged in 1992 about the 'Black Cats' and the activities of death squads. The catalyst for this debate, ironically, is a senior police officer, Premadasa Udugampola, who held a leading position in the government's military campaign against the JVP. Udugampola, after becoming persona non grata with the government, came out with a series of disclosures about political killings which he attributed to 'Black Cat' killer groups.

References

Alles, A.C., 1976. *Insurgency, 1971. an Account of the April Insurgency in Sri Lanka*. Colombo: Trade Exchange (Ceylon) Ltd.

Chandraprema, C.A., 1990. *The JVP Insurrection 1987–1989*. Colombo: Lake House.

Gunaratne, R., 1990. *A Lost Revolution*. Kandy: Institute of Fundamental Studies.

Kapferer, B., 1988. *Legends of People, Myths of State, Violence, Intolerance, and Political Culture in Sri Lanka and Australia*. Washington & London: Smithsonian Institution Press.

Kodikara, S., ed., 1989. *The Indo-Lanka Agreement of July 1987*. Colombo: University of Colombo.

Uyangoda, Jayadeva, 1989. 'The Indo-Lanka Accord of July 1987 and the State in Sri Lanka', in Kodikara, 1989.

Uyangoda, J., 1992. 'War and Peace in Sri Lanka', *Sunday Observer*, 9 February.

Vittachi, T., 1958. *The Emergency '58. The Story of Ceylon Race Riots*. London: Andre Deutsch.

9

Hindu Women: Politicization Through Communalism

TANIKA SARKAR

9.1 Introduction

A major political development in India has gone basically undocumented over the past decade. From the early 1980s, Hindu communal organizations increased the scale, range, aggressiveness and violence of their operations under the general direction of the militant Hindu right-wing party Rashtriya Swayam Sevak Sangh (RSS) and its mass fronts: the Vishwa Hindu Parishad (VHP), which coordinates religious bodies, and the Bharatiya Janata Party (BJP), its electoral wing. In recent years, this process has taken on definite shape and thrust from the *Ramjanambhoomi* movement – which mobilized the forces that destroyed the mosque built by the Mughal Emperor Babur on the alleged birthplace of the Hindu epic hero Ram at Ayodhya in Uttar Pradesh, and for the construction of a new Hindu temple there. Countrywide mobilization on the basis of the *shilanyas* (laying the foundation stone for temple building) in 1989 ushered in an unbroken spell of rioting on a scale unmatched since the holocaust of 1946–47 when India was partitioned. It led to the fall of the Janata Dal Government and the 1991 electoral success of the BJP in Uttar Pradesh.

One of the most sinister features of the recent *Hindutva* (Hindu nationalism) movement has been that militant communal Hindu women have come to the fore. Rural women at Bhagalpur in 1989, and upper middle-class women at Ahmedabad in 1990, played an active role in riot scenes – an ironic inversion of women's traditional invisibility. According to VHP reckoning[1] 20,000 *karsevikas* (women who have pledged themselves to build the temple with their own hands) courted arrest on 4 January 1991 at Ayodhya, and a total of 50,000 were involved in the entire December 1990–January 1991 round of non-violent agitation (*satyagrahas*).

The VHP fortnightly *Hindu Chetna* of 15 December 1990 displayed a cover photograph of *karsevikas* wearing the saffron headband. The caption says: 'The Rise of Mother Power'. Its Ayodhya office has been selling works by women poets which celebrate sacrifices by the mothers and wives of martyrs. An official VHP account of the October/November 1990 events (Amar Shaheed pamphlet, 1990) accords particular importance to the arrest of the leading woman political leader of the VHP, Vijayraje Scindia. Similarly, the Hindi VCR newsmagazine *Kalchakra*, covering the events of that October, focuses on the victim figure of a lone woman resister being dragged away by the police; and *Newstrack* highlights the dynamic figure of Vijayraje Scindia, who is seen striding purposefully between two senior and enormously respectful police officers towards a car. Another picture shows her inside the car, leaving messages for the future conduct of the movement. Finally, if L. K. Advani, leader of the BJP, and his *rath* (chariot) were the visual emblems of *Ramjanambhoomi*, then the voice and the words that were to fix its message belonged to a woman.[2]

9.2   Images of Hindu Communalism

The Hindu communal groundswell, it is true, has been restricted largely to the urban areas of the state of Uttar Pradesh (UP), among predominantly high-caste, middle-class milieus. However, the strength of this new phenomenon must not be underestimated. *Karsevikas* have been mobilized from traditionally the most conservative backgrounds – upper class, middle-ranking service sector and trading families. On the other hand, the very limits of the movement may be taken as signs of strength within a different kind of reading, for these women are speaking their own minds, their own words. An interviewer speaking to some male *satyagrahis* at Ayodhya in January 1991 was, for some time, faced with an array of archaeological-cum-historical arguments as well as the standard RSS definition of *Bharat* as 'Fatherland, Motherland, Holy Land and Land of Our Action'.[3] During the discussion a woman from Aligarh excitedly broke into the conversation and introduced quite a different note: 'We have come here to shed blood . . . the meaning of temple building is that *mullas* [Muslim religious leaders] should be hanged, Mulayam Singh Yadav [Congress leader] and V.P. Singh [former Prime Minister of India] should be hanged.'

The whole discussion subsequently shifted to a markedly more violent plane. This does not mean that the women were voicing mere mindless abuse. Each of the *karsevikas* interviewed – VHP as well as non-affiliated ones – played a distinctive individual variation on the themes of *Ramjanambhoomi* and *Hindutva*. For Vijay Dube, a would-be *sanyasini* (ascetic) from Ghaziabad, *Hindutva* implied a sweeping, millenarian vision of collectivity:

> It is as deep as the ocean, as endless as the sky . . . the Hindu is the beginning and the end.

This is not time-bound, but eternal; not an individual but a collective experience. It finds its centre of gravity in *Ramjanambhoomi* which then becomes: 'everything to us, it is not just a matter of religion it is all, it is our everything'. With the liberation of Ayodhya, 'the whole world will change, a new creation will come into being'.

For Mithilesh Vashisht, a VHP worker from Modinagar, on the other hand, the value of the movement lay in its assertion of strength and self-respect against oppression: 'We will not tolerate oppression and wrongs, everything has limits.' Another woman intervened with a more poetic-mythical version: 'This is a limb in our body, an ornament . . . it is the *chakra* [symbol] of Krishna *Bhagawan* [god]'.

All these *karsavikas* were bursting with speech – with arguments and descriptions, each with an accent very distinctively her own. Within a limited social and geographical scope, the *Ramjanambhoomi* movement seems to have led to major breakthroughs in women's political self-activization, unheard of in earlier communal upsurges. A relevant parallel would seem to be with phases within Gandhian mass movements, or the final stage of revolutionary terrorism in Bengal when an intensely devotional form of patriotism admitted women as fully fledged activists.

In a curious way, this movement inverts the usual pattern of symbolization within national and earlier communal movements, where the fetishized sacred or love object to be recaptured had been a feminine figure – the cow, the abducted Hindu woman, the Motherland. When B.L. Sharma, Secretary of the VHP Indraprastha unit was interviewed in April 1990, he wove an entire anti-Muslim tirade around the figure of the endlessly raped or threatened Hindu woman. Other officeholders of the party extended the image into that of a perpetually exposed and endangered Motherland.

In Ayodhya the birthplace (*Janambhoomi*) belongs specifically to a male deity. His women are being pressed into action to liberate it and restore it to him, to bring back honour to Ram. Ram's army of monkeys and squirrels has now acquired a new ally, and members of Sita's sex are coming to the rescue of Ram – a reversal of the epic narrative pattern where Ram and his army had to go and rescue Sita from her abductor. This role reversal has provided women with a new and empowering self-image. Women have stepped out of a purely iconic status to take up active positions as militants.

In this context, the cautious handling of the baby Ram image acquires new significance. Ayodhya stalls sell a large number of stickers and posters depicting a chubby infant baring his pink gums in a toothless smile. Local legend has it that in 1949, just before the deity 'miraculously' reinstalled itself within the mosque, a police constable had found a lovely, dark-skinned child playing by himself in just that corner: the homeless waif had come back home to claim his patrimony. The VHP video cassette the *Bhaye Prakat Kripala* (J. K. Jain) reproduces the event over a long time, with the child within the mosque in a variety of 'cute' poses and eventually stringing a bow. Here we should remember that Ramayan and the legendary stories of Ram resonate with the many losses of Ram: he loses both his kingdom and his father; he is separated from his mother and his brothers; then he loses his wife, Sita. His is a figure bathed in tears – one reason, perhaps, why the common man and woman can identify more readily with him than with other mythical heroes.

The entire series of deprivations have now been collapsed into the shape of that irresistible human idol – the helpless babe. Moreover, within the mosque and next to the main deity is an icon of the crawling baby Ram[4] – a posture traditionally associated with the baby Krishna and linked to a long chain of associations. While the appeal of the homeless waif would be a general one, it would be especially poignant for women. Readings of recent events that insist on a monolithic militarization of Hinduism, therefore, ignore their *versatility*, which is in fact the most remarkable feature of these events. While the baby Ram appeals to the mother instinct, the warrior Ram probably simultane-ously arouses a negative response in women devotees to an aggressive male sexuality.

All this is a relatively new development. As late as mid-1990, well after the *shilanyas* ceremonies with their attendant riots were over, there was practically no literature written by women that was sold regularly by VHP or RSS offices. VHP newsheets gave no space to women's writings. Saddhi Rithambhara's audio cassette addresses its invocations to rise to fight exclu-sively to men: 'Brave brothers, wake up'. Her speech does target women listeners as well, with close references to domestic politics among mothers, sis-ters and daughters-in-law, to women's work within the home. Yet whenever the call for action is issued, it is addressed to brothers: 'You have to make yourselves into a clenched fist, my brothers.'

Inspirational feminine examples relate to motherhood: how the mother of Bhagat Singh was found crying after his death, not because she had lost her son, but because she had no other son to be martyred. Even the Rani of Jhansi, a Muslim, is invoked as the mother of a brave patriot. It is true that the *Bhaye Prakat Kripala* cassette inserts the warrior figure of a queen as an adversary of Babur; and, within the current movement, women leaders like Saddhi Rithambhara, Satyavani and Vijayraje Scindia are endowed with an exalted position. These are, however, exceptional figures. Up to this point, women have still been the productive womb, the mothers of heroes. Their presence was minimal in the earlier crucial days of the *Ramjanambhoomi* movement. At the Rashtrasevika Samiti office it was quite frankly said that the decision to train *karsevikas* was the result of an internal debate that was eventually won by younger Samiti members.

9.3   The Female Face

A new shift has occurred very recently, a shift full of possibilities – and prob-lems – for Hindu communalism. Communal organizations have not allowed a demonstratively public or even a very active political role to women so far. The RSS is an exclusively male organization. When in 1936, Lakshmibai Kelkar approached Hegdewar (an RSS leader) with an appeal to admit women as members, she was refused. Hegdewar later helped her to set up a parallel but separate organization – the Rashtrasevika Samiti. There have been debates within the RSS about membership for women, but so far it has remained uncompromisingly male.

The Rashtrasevika Samiti has maintained a remarkably low public profile through the six decades of its existence. One of the oldest women's organizations in the country, its total membership is only about 0.1 million now, largely restricted to traditional RSS and BJP bases – Maharashtra, Karnatak, Andhra Pradesh. The Delhi wing, formed in 1960, now numbers about 2,000 members. Training programmes are located almost entirely in middle-class areas. Volunteers come from enterprising trading families or from middle-ranking government service backgrounds.

The VHP Mahila Mandal (women's wing), which started operating in Delhi during the 1980s, has about 500 members. The two mass fronts of the VHP – the Bajrang Dal and the Durga Vahini – are strictly segregated. Growth in spatial, numerical and social terms has been quite low compared to the Delhi-based radical women's organizations under Left political parties. The All India Democratic Women's Association, linked to the CPI(M), was founded in 1981. It now has a membership of about 2.9 million, overwhelmingly rural in composition. Its Delhi branch has about 15,000 members, with large bases in working-class areas. Comparison here is relevant since the VHP Mahila Mandal and the Rashtrasevika Samiti (unlike other radical voluntary women's organizations) also work in close collaboration with a political party and a number of affiliated mass organizations.

How can we interpret the apparent contrast between the austere reserve of the organizations and the recent flamboyant wave of a militant reclaiming of the public spaces by women? Is it a break, a total departure – or is it the culmination of a long, drawn-out strategy? Or is it perhaps a consciously planned extension that stretches out old boundaries?

Too often, notions about social conservatism are imputed to the *Hindutva* movement without considering the gaps, even contradictions, between the two. That is, we tend to read the new *Hindutva* movement as but another reiteration of the gender ideology of traditional Hinduism, of the sort that surfaced during the spate of *sati* (self-immolation of widows) in Rajasthan in 1987. There are, obviously, certain resemblances between the two. The Hindu right-wing evokes Hindu scriptures as the final court of appeal. The RSS is an all-male body with a marked accent on the traditional concept of celibacy. Leading BJP spokespersons like Vijayraje Scindia defended the practice of *sati*. Yet, it is the differences that are more interesting, as they are ultimately far more revealing of the directions which the new Hindu woman is going to follow, and of the problems and contradictions that she may have to face.

The Rashtrasevika Samiti was, as noted, created by the RSS. The question of the autonomy and independent policies is crucial – also for assessing how far the purely female Samiti relates to the purely male Sangh. The gender question is a key to understanding the *Hindutva* movement as a whole.

An intricate and delicately balanced system of interlocking personnel, functions and interests is ultimately monitored by the apex parent organization – the RSS. Indeed, the Sangh calls itself a 'family', not a political organization, and claims that all its members are equal and uniform in dress, disposition and functions. This, however, must stem from a peculiar notion of a 'family'. There cannot be a family without women in it, nor can there be a family that is undifferentiated in functions and habits: thus, only by developing a women's wing does the RSS family metaphor partly realize itself. The Rashtrasevika Samiti was the first affiliate that the RSS helped to foster, a good 11 years after its own foundation. Nor is the family model merely metaphoric: all the Samiti members interviewed had male relatives in the RSS. In fact the striking ease and self-confidence that animate the highly vocal participation of even junior officeholders in discussions with their elders may partly be explained by the status of their male relatives within the Sangh, which might carry greater importance than the order of ranking within the Samiti itself. Within the Delhi VHP Mahila Mandal, too, the three top figures are all married to VHP leaders.

The organizational principles of the RSS provide the pattern for the Samiti. It does not have internal elections; the head of the organization nominates her own successor. Officeholders are selected by senior members. *Samiti pracharikas* (teachers/trainers) are unmarried women, echoing the RSS accent on celibacy. Much of RSS ritual is also replicated within the Samiti. Common to both are the strict and detailed code of regulations for daily *Sakha* training programmes, the training for physical and martial arts (the Samiti provides for lessons in yoga, sword and *lathi* play, judo as well as shooting), regular or ideological discussion (*boudhik*), and the overall discipline and protocol.

Whereas the RSS observes six annual festivals – five of which are connected with traditional religious events and the sixth, the Shivaji Utsav, is a Hindu nationalist celebration – the Samiti observes five and omits the non-traditional festival in honour of god Shiva. The major hymns are also more or less the same and are recited individually as well as collectively in exactly the same order. At the same time, a fine tension exists between vociferous claims to complete autonomy and a pride in sharing the RSS heritage. Viduhshi, a young office-bearer of the Samiti, told me rather defiantly that the two bodies are totally distinct, like two railway lines which run parallel, yet are always separated. This analogy, however, comes straight from Golwalker, the founder of RSS (1925). In a different sense (discussed later), the Samiti supplements certain kinds of Sangh work, in a way reiterating the conventional place of the Hindu righteous wife, *dharampatni*, within the household – a related yet subordinate sphere.

The sessions for ideological discussion (*boudhik*) within the Samiti repeat the same finished structures of thought, the same basic themes on which variations are played by the RSS, the BJP, the VHP, individual male and female workers and loose fragments, which have come to constitute parts of popular commonsense among non-affiliated informal support groups for a Hindu state.

9.4 Evolving Identity

Is the Samiti then merely filling up a space marked out for it by the Sangh for its own purpose? And, if so, would not this contradict a statement made earlier here – that the strength of the movement lies in the exhilarating possibilities it provides for certain sections of deeply conservative women, in its being an expression partly of their own creativity?

In my view, the original parameters worked out by RSS have proved to be flexible and accommodating. The major body of Golwalker's instructions for women had set up an image of faithful motherhood within which the properly instructed mother carefully guards her children from corrupting Western influence and instils in them the right principles – filial piety, knowledge of patriotic heroes and of religious texts. Members of the Rashtrasevika Samiti have travelled a long way since then, without overtly contradicting the original instructions. The gap between the original impulse and the new self-definition may or may not open up fissures within the movement, depending upon the versatility and suppleness of Sangh ideology – both of which seem considerable.

Somewhat different pulls appear to be working within the Samiti itself, though it would be an oversimplification to categorize them as potentially feminist versus overtly fundamentalist ones. Varieties of emphasis, none the less, remain interesting. In the first place, there are somewhat different versions explaining the origin of the organization. Neither the authorized history of the Samiti, nor the accounts of a senior officeholder and an activist mentioned anywhere that Hegdewar was first approached to admit women into the Sangh and that he refused and helped Lakshmibai Kelkar to set up a parallel organization instead. It is an RSS publication that dwells on this fact and claims that Hegdewar convinced Mrs Kelkar about the difficulties that having a common organization would involve. The official Samiti history, however, has a somewhat different explanation for the Samiti's foundation:

> With the primary aim of the awakening of *Hindutva*, this independent organization began its auspicious start.

Rekha Raje, a leader/trainer, however, has narrated how Mrs Kelkar on a train-journey had witnessed a young girl being raped by ruffians (not Muslim ones, she replied to my question) in her husband's presence. She realized that since Hindu husbands cannot protect their wives, the women would have to strengthen themselves. Whereas the edge in this story is turned against failures within Hindu society – Hindu male violent lust and Hindu male cowardice – the official text overlays that version with the broader and more general aim of *Hindutva*-awakening.

The self-explanations and self-definitions produced by the Samiti emphasize physical courage and strength, and the trained, hardened, invincible female body. The Sangh, too, defines this – cultivation of the physical strength of Hindu women – as the first principle, and then goes on to list 'intellectual grasp of the values of Hindu culture and devotional attachment to the ideals of Hindu womanhood'. A Samiti publication puts it in stronger terms: 'a woman who is able to defend herself gets a higher status in society'.

The specific deity which sums up their aspirations is the militant icon of

Durga, who subsumes such other female goddesses as Saraswati, Lakshmi and Kali. Members of the Samiti are meant to meditate on Durga's weapons particularly. They see themselves as fully fledged soldiers in an impending apocalyptic war; their daily pre-meal mantra is:

> Our limbs and bodies have been nurtured by our Motherland and we must give them back to her in her service alone.

Why is a strong female body of such primary importance? Expounding on this theme, Asha Sharma, who is in charge of the Delhi organization, explained that this binds up the notion of sacrifice with that of active fighting. When asked if it was some kind of a civil war situation that she had in mind, she replied that this was a possibility. Certainly then, the explicit purpose for which the empowered Hindu female body is trained is a 'patriotic' war against the Muslim combatant. The considerable space allocated to the myth of Muslim lust within the general mythology of Hindu communalism would also explain the need for physical self-strengthening. Yet we should remember the oral version of the origin myth – Hindu criminals raping a girl in the presence of a Hindu husband – and also the reference to the higher status of 'a woman also can defend herself' within her own social milieu. Thus, defence against and respect within her own environment is the implicit subtext which might well become a more powerful motive force and a more vital compulsion than the ultimate political intention of *Hindutva*-awakening.

When we understand the context from within which the Samiti mobilizes and trains its women, the force of the immediate compulsion becomes clearer. Members come from upwardly mobile, urban, solvent, trading or middle-ranking service sectors – a fertile breeding ground for dowry murders and the violence against wives that precedes them. Women's organizations, which deal with huge numbers of divorce or maintenance suits at this social level, are all too familiar with the flourishing violence and oppression against women. In the big northern cities, education and professional opportunities for women may have been late in coming, but now they have arrived in a fairly big way. Nor are families opposed to women's employment and professional training, as these are regarded as a valuable source of extra income. Thrust into public and mixed-gender spaces for the first time, women daily encounter new forms of overt or covert sexual discrimination and violence. It is no wonder that the physical training programmes of the *Sakhas* prove extremely attractive to such women, with the promise of a powerful body and the attendant self-confidence. The physical and mental attitude thus generated could serve as a vital shield against gender oppression within domestic as well as public spaces.

Thus, we may assume that despite the overarching aim of Hindu power, women also need to utilize Samiti facilities to empower themselves against their own hostile environment. Problems of newly mobile professional women are often discussed in *Jagriti*, the Samiti journal. One article, for instance, describes how the author withstood the offensive behaviour of a police officer who partially undressed himself in her presence when she had gone in to report a street accident. Several others take up this theme: how to construct a responsible and fearless woman-citizen and teach her to exercise her civic rights and duties within a chauvinistic world.

Two very different readings of the precise location of the Hindu woman within her own society seem to be jostling with each other within the Samiti. The speech by the Samiti head at the 1990 annual conference insisted that 'the women of India have always been free'. An article in *Jagriti* reiterates this conviction and traces a long history of women's power within Hindu society from Manu (saint and first codifier of Hindu law) to the current movement. Yet that same issue carries another article in which the present women's movement in India (which is not equated with the specifically Hindu movement) is described as an antidote to the generalized oppression against women. Whereas the first approach is extremely critical of the global women's movement as a sign of 'Western corruption', this second article explains and legitimizes it. Another article criticizes Indian men for obstructing the overall entry of women into politics.

It is interesting to see that while the more authoritative statements – Golwalker, RSS structures, official Samiti accounts – applaud the new Hindu woman for resisting Western modernism, the women's own articles, in dealing with their everyday problems and perceptions, show scant concern about Westernized modernity. In fact the new Hindu woman citizen is cast in a mould very close to that of bourgeois feminism:

> In order to attain the comprehensive development of women it is extremely important for them to be economically independent. So, in order to ensure economic independence, they need reservation in employment and they need women judges to conduct all cases related to such issues. (*Jagriti*)

The 'new' Hindu woman is thus seen as a person with professional and economic opportunities, secure property ownership, legal rights to ensure them and some amount of political power to enforce these rights.

How does this woman relate to Hindu tradition? Hindu tradition would seem to exist for her as a deeply cherished yet somewhat remote icon, one which requires ritual worship but is seldom brought out for daily use or inspection. There is very little interest in probing it and deciding what it actually says about gender. In fact, the question of religious faith is often interpreted as completely displaced onto the realm of patriotic faith. The greatest triumph of today's communal movement has been its success in blending devotion to the country with devotion to Ram – the two potent sources of emotional involvement – into a homogenized whole so that adherents use them interchangeably. When our interviewer asked the *karsevika* Vijay Dube, a woman who plans to adopt *sanyas* (asceticism), how *Rambhakti* (devotion to Ram) has come to mean so much to her, she immediately traced back its source to her childhood experience of the China War and the passions it had aroused within her. Patriotic faith, in her case, was the original impulse that stoked religious passion, and she herself was not aware of any distinction between the two. The process of mutual collapsing can work with equal felicity the other way round as well. At Ayodhya, it was the banning of the usual pilgrimage route that quickly turned pilgrims into *karsevikas*.

On specific questions of gender within Hinduism, the Samiti verdict seems unequivocally modernistic. *Sati* was squarely denounced by all officeholders. When I asked about voluntary *sati*, a young activist said with genuine revulsion: 'It isn't possible. Why should a woman burn?' She then went on to explain that she herself might want to do it out of 'depression and frustration' that is, as a mark of weakness, not as a mark of strength – but that neighbours and relatives should shore up her will to live.

In theory, the Samiti does not ban inter-caste or even inter-communal marriage —provided, of course, that the family agrees. The visual imagery presented on a recent *Jagriti* cover is clear. It depicts two passive women against a black background, crouched in a posture of helpless mourning. And out of that dark frame, a young, rather grim-faced woman is stepping out onto the radiant half of the cover with a firm stride and uplifted head. She bears no traditional Hindu marks – no *sindur* (red colour in her hair), no veil, no *bindi* (coloured mark on her forehead). She is wearing *chappals* (slippers), her *sari* is draped securely around her and her whole stance is free, even aggressive. Of course, she is wearing the Samiti uniform – the purple-bordered white *sari*. A striking feature of this magazine is its uncompromisingly non-feminine nature. It makes no concessions to conventional women's topics. There are no hints on beauty aids, no cookery or embroidery sections, no advice on child rearing. Stories are scrupulously bare of the romantic element, dealing instead with the romance of Hindu civilization or that of modern patriotism.

Women's power is a theme that is evoked in the Samiti and celebrated, often in the most grotesque of circumstances. At the *Ayodhya karseva*, a crowd of women stood chanting the lovely feminist slogan: 'We are the women of India. We are not flowers, we are sparks.' When our interviewer asked them why Sita was absent in their invocations to Ram, the men fell silent but women had their answers ready. One said that this was Ram's birthplace and not Sita's, which accounts for her absence. But Vijay Dube interrupted her to say that in the chant 'Shri Ram', 'Shri' actually means Sita; hence Sita is actually placed before Ram. The interviewer asked – 'You mean Sita is contained in Ram?' 'No,' said Dube, 'Sita comes before Ram.' Not only had thought gone into the location of Sita, there was also a recognition that Sita must come before Ram. In a VHP book for children – *Hanuman ki Kabaddi* – Hanuman (the monkey god) declares himself to be neither Ram's devotee, nor Sita's, but the follower of Sitaram together.

The new Hindu woman is sombre, efficient, self-sufficient and mobile. She is a citizen, a professional woman with her own property rights, a mother and a homemaker. She is emphatically not an erotic figure but proudly owns a hard, strong body. She is not a figure of romance but a severely serious person.

At the same time, within her home and professional world and in her organizational capacity, she is required to follow dictates, and accept the morality
of a hierarchical, unequal social world. She is not to set up oppositional sol-
idarities along lines of caste, class or gender; and while she is expected to
work, she is not expected to be a union member.

This model gives rise to expectations among the broad political milieu of
new *Hindutva* where domestic and social conventions remain too restrictive
for most women to realize the pattern in their own lives. At Khurja, a small
town in Western UP, female members of an RSS leader's family told us that
domestic responsibilities stifle their political interest and educational hopes.
The sister-in-law of a dynamic young girl who had tried to set up a Samiti
branch there remarked ironically: – 'She did a lot and hopes to do a lot, but
her days are numbered and her "number is coming up"' – that is, marriage is
approaching.

9.5   Gender and Class

I do not wish to convey the impression that some sort of women's liberation is
happily going on within a rather unfortunate *Hindutva* framework. When I
asked Asha Sharma where the Samiti differs from other women's organizations,
her reply was immediate:

> We do not believe that in marital disputes the husband is necessarily to blame.
> When we arbitrate, we do not take the woman's side, we are neutral. We tell the
> woman that she must do everything to preserve her home life. We are not wreckers
> of homes.

The Samiti offers no formal legal counselling to women, nor is divorce generally encouraged. Dowry is regarded as an evil, yet there obviously is no ban
against its practice among Sangh or Samiti members. If demands for full citizenship rights and affirmative action are made in *Jagriti* from time to time,
there is no critical review of Hindu patriarchy. When *karsevikas* at Ayodhya
were asked if their status would improve within a Hindu state, one of them
said yes – because Muslims then would not be allowed to have four wives and
that alone would ensure greater respect for women. However, she could not,
on the spur of the moment, think of any other possibility within *Hindutva* for
herself.

Just before the mid-term elections in May 1991, some women activists
publicly interrogated spokespersons from major political parties about their
programmes for women. The BJP spokesman was most often at a loss and
found very little to say to the women. He also infuriated many working-class
women from slum areas by chastising them for bringing their babies into the
meeting hall. 'Are our babies shoes and slippers that we can leave them
outside?', they retorted.

Caste is formally denounced and Sakha members are not to submit their
surnames when they enrol, so that their caste status does not become public

knowledge. There is also community dining to ensure the absence of barriers. On the other hand, the caste system is not a theme for discussion or criticism in *boudhik* study sessions. As with gender, class struggles are yet another field of resounding silence and non-involvement in general.

9.6   Conclusions

Where do feminists stand in relation to this manifestation of women's power in India? There is no denying that it does empower a specific and socially crucial group of middle-class women – if not in an absolute feminist direction, then definitely in a relative sense. It helps hitherto homebound women to reclaim public spaces, to acquire a public identity; it confers upon them a political role and even leadership. It teaches women not to regard themselves as merely feminine but as fully fledged citizens. It gives them access to serious intellectual reflection.

The costs too are obvious. The public identity is regimented, colourless, even grim. Intellectual discussion and political involvement have culminated in a violent campaign of blind hatred, instead of in a critique of class, caste and patriarchy. The movement prepares women for citizenship in an authoritarian Hindu state, and to reject secular, democratic politics.

This paradox of self-limiting but violent self-empowerment cannot be subsumed within currently fashionable, flat and facile critiques of post-Enlightenment modernity. It reads *Hindutva* as yet another problematic manifestation of a necessarily distorted and Westernized discourse, and seeks resolution within so-called unproblematic, accommodating, warm and generous Hindu tradition, in a shift from modernistic *Hindutva* to traditional Hinduism. Socialists and feminists can hope for no resolution for issues of caste, class or gender within tradition, nor can they afford the luxury of ascribing all present distortions to a simplified, flat notion of colonial discourse. In any case, the specifically modernistic elements in the communalism of Hindu women are hardly ones that they can identify as sources of distortion, or recommend their abolition – the Samiti's verdict on *sati* being an obvious example here.

The real problems would seem to be in the realm of active politics. Resolution, then, may be found in more aware and sensitive forms of Left democratic and feminist movements.

Notes

1. This was conveyed to the author and her team at the VHP's office at Delhi in February, 1991.

2. In October 1990 L.K. Advani undertook a spectacular all-India tour on a chariot driven by a Maruti engine which was supposed to culminate at Ayodhya and inaugurate temple-building. He was, however, arrested in Bihar; his party subsequently withdrew its support from the government and brought it down. The whole journey blazed the trail for violent anti-Muslim rioting throughout the country.

3. *Pitribhumi, Matribhumi, Punyabhumi* and *Karambhumi.*

4. In 1949 an icon was installed within the mosque by a group of Hindus. A controversial court decision allowed the icon to remain there.

Bibliography

Amar Shaheed pamphlet, 1990, an official VHP account of the October/November 1990 events.
Hindu Chetna, 15 December 1990 (VHP fortnightly).
Jagriti, the Samiti journal.
Kalchakra, the Hindi VCR newsmagazine, covering October events.
Saddhi Rithambhara's audio cassette.
Sharma, B.L, 1990. Interview with Sharma, the Secretary of the VHP Indraprastha unit.
VHP video cassette, *Bhaye Prakat Kripala* (J.K. Jain).

10

The Gender Dimension in Sindh's Ethnic Conflict

KHAWAR MUMTAZ

10.1 Introduction

Women's participation in movements and struggles is not uncommon. This is evident from the active involvement of women in many struggles of this century – in Algeria, China, Iran, Vietnam, to mention a few. However, it appears that the nature and extent of women's involvement is determined by the conjunction of several factors – economic, political, cultural – at a specific point in history. Thus, the anti-colonial struggle in the Indian sub-continent attracted women, whereas in post-Independence Pakistan neither the agitations for greater autonomy in North West Frontier Province, the secessionist movement in East Pakistan, nor the insurgency in Baluchistan saw the organized participation of women. On the other hand, the ongoing ethnic conflict in Sindh – between the indigenous people of Sindh and *mohajirs* (Urdu-speaking immigrants who came from India at Independence in 1947) – has seen large-scale mobilization among the women of the two communities involved.

In fact the 1980s in Pakistan witnessed the growing political activization of women at two levels. One in response to the politicization of gender, which, according to Valentine Moghadem, occurs 'during periods of transition and restructuring when social groups and values clash', especially in 'patriarchal societies undergoing development and social change' (1992, p. 49). The other, in support of the national ethnic cause. The former represents women's mobilization to safeguard their rights when these are threatened by state-initiated measures; the latter is integral to the movements of the two major ethnic communities of Sindh.

Women's organized resistance to the discriminatory measures of General Zia-ul-Haq's dictatorial regime has had a significant impact on women's groups and organizations as well as political parties – a fact widely, if grudgingly, acknowledged in Pakistan. Not only has the women's issue been placed on the agenda of political discourse: also political parties – including those of the religious Right – have felt constrained to set up women's branches. Whereas the autonomous movement for women's due position in society (spearheaded by Women's Action Forum) and the debate on the 'woman question' it triggered have been documented and analysed (Rouse, 1986; Mumtaz & Shaheed, 1987; Gardezi, 1990; Shaheed & Mumtaz, 1990; and others), the activism of women from ethnically defined groups currently in conflict has not been studied.

This chapter seeks to examine the dynamics of women's induction into the Sindhi nationalist movement and that of the Mohajir National Movement (Mohajir Qaumi Mahaz). The former is not a unified entity but a composite of various parties and groups representing a range of positions within the broader framework of Sindhi nationalism, from those seeking greater autonomy to the secessionists. The chapter will focus on the MQM women's wing and the Sindhiani Tehrik (Sindhi Women's Movement) – the women's wing of the Left party seeking greater autonomy within the federation of Pakistan, and the most organized among women in Sindh.

Today's ethnic conflicts are new issues arising from new social dynamics. Since 'ethnic consciousness is less an inheritance from ancient times and more sharpened and heightened ideology created by recent, sometimes bitter events' (David & Kadirgamar, 1989, p. 13), it is indeed relevant to examine the nature of women's involvement in the Sindh conflict. What are the roles that women are envisaged to play? Do women's issues feature in, or are they subordinated to, the larger goals of the ethnic movement? Do other women's movements (e.g. that led by WAF or that of the religious right countering WAF and defining women's rights with reference to Islam) have any implications for them? Does the participation of women reduce the violence in these movements? With women's involvement, do more creative ways of conflict resolution evolve?

These and other related questions need answers, possible only after an exploration of the two groups. This chapter is a preliminary effort in that direction.

10.2 Political Backdrop

Pakistan is among the post-colonial societies where ethnicity is 'an internal phenomenon' with features similar to those so succinctly described by David & Kadirgamar (1989). According to them, conflict emerged in such societies when attempts were made to impose a modern, integrated, 'bourgeois' state from the top, on to 'smaller self-identifying communities'. In the case of Pakistan, the fact of its pre-capitalist stage of development and diverse social formation and cultures (essentially feudal) on to which the super-imposition of a bourgeois nation-state was sought, provided 'the necessary material conditions . . . for regimes of crisis, for ethnic conflict and . . . for the collapse of the whole project of bourgeois nation-building' (1989, pp. 2–3).

The specificities and peculiarities of this process of collapse are dealt with and analysed in some detail elsewhere in this volume (see Hussain (Chapter 3), Rashid (Chapter 4), Mazari (Chapter 6)); suffice here to say that the tussle between the military and representative institutions has been endemic to the process. This has taken place in complete disregard of people's needs, their expectations or aspirations. And, in the absence of the development of a democratic process, ethnic identities have become increasingly important in deciding the parameters of people's lives.

Over the years, beginning with the martial law imposed in 1958 by General Muhammad Ayub Khan, the process of transforming the Pakistani state into a despotic one culminated with General Zia's military takeover in July 1977. Religion, first used by ruling elites as a unifying ideology, became the tool of emerging political groups for their mobilization. Zia's military coup marked the juncture at which the worldview of the obscurantists coincided with that of the state.

The consequences of such a 'flawed' political process have indeed been far-reaching:

- failure to develop a pluralistic system to cope with the needs of the multiple identities of the people within the country;
- religious and ethnic identities gaining fundamental significance in defining people's lives;
- religion, a powerful symbol of national identity, turning into a divisive medium and becoming a tool for legitimizing unpopular, authoritarian regimes;
- the deepening crisis of identity of Pakistan's people and the formation of ethnic organizations alongside the emergence of a host of religion-based ones;
- the militarization of the state and large-scale introduction of modern sophisticated arms (spurred by the Afghan War and the concomitant drug trade) adding militancy, unprecedented violence and brutality to contemporary conflicts.

10.3   The Conflict in Sindh

10.3.1   Underlying Causes

In Sindh, government policies have resulted in a deep sense of alienation among the Sindhis, who have felt discriminated against and wronged almost since Independence in 1947. Their premier city, Karachi, was taken over by the Centre to be made the capital of the new country; their land (almost 40% of Sindh's land area) mortgaged to Hindu moneylenders was not returned to the original owners (although similar land in Punjab was) but declared evacuee property and allotted to Urdu-speaking immigrants; additional land colonized through the building of barrages was allotted to senior civil and military bureaucrats and non-Sindhis;[1] and their language (Sindhi), the most developed of Pakistan's regional languages, was in 1958 abolished as the medium of instruction in educational institutions. Hence the strong feeling among Sindhis that they have been denied their due share: in jobs, in educational institutions, in politics. Sindhis resent the fact that their warm welcome of the Urdu-speaking immigrants was taken for granted. That there is some basis to their various complaints seems evident:

In no other province have outsiders been chief secretary and inspector general of police for 43 out of 45 years. In none other was the chief executive an outsider for thirty out of forty-five years. No province is ethnically more integrated, or economically more dominated by outsiders as Sindh. No ethnic group in Pakistan has as negligible a representation in the armed forces as Sindh. (Ahmad, 1992)

The course of political events contributed further to a Sindhi sense of disillusionment with the Centre and with Pakistani nationhood. In 1955, promulgation of 'One Unit' – whereby all provinces in the Western wing of the country were merged into one; the imposition of martial law to overthrow the government of Prime Minister Z.A. Bhutto in 1977; his farcical trial and execution (1979); the hounding of Pakistan People's Party activists; and later the ruthless crushing of the rural-based movement launched in Sindh for the restoration of democracy (MRD) in 1983 – all have been perceived as measures directed against the people of Sindh. The fact that urban inhabitants of Sindh did not support the MRD (nor did the other provinces) exacerbated feelings of isolation and alienation.

On the other side, the Urdu-speaking *mohajirs* have progressively also experienced a sense of alienation and isolation. Two-thirds of the *mohajirs* are settled in urban areas of Sindh. Filling the vacuum left by fleeing Hindus, they replaced the Hindu middle class, taking over trade and commerce. Better educated and trained than the indigenous Sindhis (who at Independence comprised the peasantry and landed feudals), they dominated the bureaucracy and the armed forces along with the Punjabis. Feeling superior about their language and culture, they made little attempt at integration with the Sindhis, and instead identified with the Centre. As a result, when 'One Unit' was broken up before the 1970 general elections, they were resentful at the return of Karachi to Sindh – the demand for Karachi as a separate province for *mohajirs* has been raised from time to time since 1962. When Bhutto reintroduced Sindhi as the official language of the province, this left them threatened, and Bhutto's initiatives (1973) to address Sindh's grievances through allocation of urban-rural quotas for government service and admission to educational institutions created a sense of deprivation.[2]

Over the years, Sindh experienced lopsided development, with little investment in rural areas despite urban industrialization. This led to the rapid influx of migrants from other parts of the country into Sindh's urban areas, particularly Karachi. At the same time, the shifting of the capital from Karachi to Islamabad in the 1960s created a 'vacuum of power' in the city (Rashid, 1991, p. 112) contributing to the decline of *mohajirs* in elite decision-making positions. As new classes emerged in Sindh (rural middle class; urban, educated unemployed), and older ones (feudals) came to manipulate more political privileges, *mohajir* frustration increased. A dichotomous situation thus emerged in Sindh: 'a rural economy and an urban economy – one feudal, the other capitalist; one Sindhi, the other non-Sindhi' (Ahmad, 1992).

While this situation has its roots in history as well as in cultural and economic development, the responsibility for the current state of the crisis is generally laid at the door of General Zia-ul-Haq. Not only did he bring about a prolonged confrontation with the people in his obsession with routing the PPP, he overtly and covertly supported the ethnic groups in both rural and urban Sindh. On the one hand he is viewed as having encouraged the creation of MQM in urban Sindh, on the other of having supported the extremist Jiye Sindh Mahaz ('Long live Sindh' Front) in the Sindh interior. The Punjabi Pakhtoon Ittehad, an alliance of Punjabis and Pathans in Karachi, also came into being at this time.

The polarization continued after Zia's sudden death in an aircrash in August 1988 and crystallized over the following years. In 1988, when Benazir Bhutto's PPP came to power in Sindh and the Centre, the province saw some of the bloodiest clashes between the PPP and the MQM. The toppling of her government and the coming to power of the Islami Jamhoori Ittehad (Islamic Democratic Alliance) in 1990 witnessed a pogrom conducted against the PPP, followed by the launching of the army's Operation Clean Up against the MQM.

Despite the clipping of the MQM's wings by the army, the ethnic divide has continued to grow. The province, ridden with bitterness and violence, has been the scene of waves of unprecedented crime that keep erupting with increasing ferocity.[3] The authority and legitimacy of the state has been repeatedly challenged with accusations of being a partner in the ongoing violence and bloodshed. As one commentator put it:

> Nowhere has the state sanctification of violence been so pronounced as in Sindh, and nowhere have the wages of arms, drugs, ethnicity, deprivation and despair been so great. Sindh's destiny seems to have been written in blood, chapter after gory chapter, by the custodians of the law, by the usurpers and wielders of power, and the final tragedy, by those of its own soil. (*Newsline*, October 1991, p. 9)

> Karachi's killing fields have been activated again, and its hapless citizens are reaping the harvest. The city's latest bout with violence has proved once again, if proof were needed, the ham-handedness of the army, the short-sightedness of the government, the armed might of the MQM, and the havoc that can be wreaked by a combination of the three. (*Newsline*, May 1994, p. 11)

Rampant lawlessness, the free movement of dacoits, supported and abetted by the administration, feudals and politicians, kidnappings and ransoms have been the source of terrorization of ordinary citizens and threaten to continue for some time to come. Ethnic lines have split the administration, educational institutions, hospitals, trade unions and neighbourhoods; even intellectuals and journalists. In this highly polarized environment, the saner elements from both sides – liberals and pacifists – have been sidelined and forced into silence.

10.3.2   The Actors

The major contestants in Sindh's ethnic conflict are the Sindhi nationalists and the *mohajirs*. The latter are represented by the Mohajir Qaumi Movement (MQM); the former by a number of Sindhi groups and parties from wide-ranging ideological moorings and their various subsidiaries and breakaway factions.

Sindhi Nationalists   Sindhi nationalist assertion dates back to the opposition to One Unit in the 1950s. One Unit was an attempt made by the ruling elite to gloss over the differences between the various nationalities/sub-nationalities that existed in the country, and to deny the Bengalis of East Pakistan their numerical superiority. Today, we can discern four streams among Sindhi nationalists:

- those seeking secession and an independent status for Sindh – *Sindhu Desh* (land of the Sindhis);
- those believing in a confederation;
- those demanding enhanced autonomy in the federation of Pakistan;
- those who desire a more just dispensation within the existing federal arrangement.

None of the above, however, is represented by one party or group. Each in fact has a number of exponents, often pitched against each other.

Jiye Sindh Mahaz (JSM) is one of the organizations that grew out of the anti-One Unit united front. Formed in the late 1960s in the closing years of Ayub Khan's martial law, it represents the extreme faction. It is the political front of the Jiye Sindh Supreme Council, and includes students, women and labour fronts.[4]

G.M. Syed, octogenarian politician and founder of JSM, who was in the forefront of the Pakistan movement in the late 1930s and in the 1940s, is the acknowledged father of Sindhi nationalism. He now says: 'We are not at all interested in democracy because democracy is the name of the 66% Punjab majority. We consider such democracy as poisonous for Sindh' (Mirza, 1986, p. 72). Advocating independence for Sindh, he has been in and out of prison on various charges: working against Pakistan, desecrating the national flag and for raising the *Sindhu Desh* flag in Karachi on 17 January 1992, at a public meeting celebrating his 89th birthday (*Frontier Post*, 18 January 1992).

A *pir* (spiritual leader), G.M. Syed belongs to the landed class of Sindh. He reaped benefits from Zia-ul-Haq's patronage, who for political expediency promoted Syed as a counterpoise to Benazir Bhutto and the PPP in Sindh. But he is respected for his personal behaviour and attitude to the cadres. Representing feudal interests, G.M. Syed's demand for *Sindhu Desh* does not put forward any social or economic restructuring. In an interview in *The Herald* (Abbas, 1989, p. 183) he goes on record as saying: 'certain communists . . . insist I first settle what the system will be. I have definitely debarred these questions before we achieve our independence, because I smell a rat.'

In 1988 the JSM attempted to broaden its base by forming the Sindh National Alliance. The SNA put up candidates for the 1988 general elections, but even this broader alliance could not win a single seat in the face of the PPP wave in Sindh (Abbas, 1989, p. 171).

The JSM student wing, the militant Jiye Sindh Students' Federation (JSSF), became more important than the JSM. It controlled campuses in Larkana, Karachi, Nawabshah, Sanghar, Hyderabad and other Sindh towns. Armed with sophisticated weapons, the JSSF was held responsible for some of the armed attacks on *mohajirs* in Sindh, and was said to be linked with the dacoits in Sindh. One of its leaders, Dr Qadir Magsi, in prison until November 1993 for allegedly masterminding the attack on 200 *mohajirs* in Hyderabad in 1988, leads his own faction, the Jiye Sindh Tarraqi Pasand Party. He dissociated himself from G.M. Syed in January 1992, around Syed's birthday celebrations (*Frontier Post*, 14 January 1992). Various other leaders within the JSSF represent different factions in the party.

Rasul Bux Paleejo's Awami Tehrik defines itself as a left socialist party. It is better organized than the others, and has students' (Sindh Shagrid Tehrik) as well as women's (Sindhiani Tehrik) wings. It is declaredly anti-feudal and vehemently anti-*mohajir*. It represents the emerging Sindhi petty bourgeoisie and small proprietor-farmers. Awami Tehrik demands more autonomy for the provinces in Pakistan and the restoration of civil liberties for all citizens of the country. The party has limited influence, being strong in two divisions of Sindh – Thatta and Badin – and is expanding to Larkana. Awami Tehrik was a member of the SNA formed in 1988. Its founder, Paleejo, is viewed with suspicion by the JSM. G.M. Syed considers Paleejo an 'agent of the CIA and an agent of the Punjabi people' (*The Herald*, August 1989, p. 182).

Other strands (e.g. Sindh National Front; the confederation-seeking Sindh–Baluch–Pakhtoon Federation; Communist Party of Sindh), as well as various attempts at forming united fronts and loose alliances, have had a limited and localized impact. Most of these have well-armed student fronts, usually more active and militant than the parties themselves.[5]

The Pakistan Peoples Party plays the very important role of stemming extremist tendencies in Sindhi nationalism. It, however, found itself caught between conflicting pressures of Sindhi extremists and *mohajirs* in its first period of political power (December 1988 to August 1990). The Sindhis viewed the PPP government as betraying the cause of Sindh and not living up to their expectations; the *mohajirs* in turn accused it of undermining *their* hopes.

Despite differences in their proclaimed positions, all Sindhi nationalists are united on a number of issues: in their perception of the Centre's discrimination against the Sindh; their strong sense of alienation from, and suspicion of, the Centre and the army, Punjabi politicians and bureaucracy; and deep-seated anti-*mohajir* feelings verging on chauvinism. There is unanimity in opposing the construction of the Kalabagh Dam on the upper reaches of the River Indus;[6] concerning the repatriation of Biharis from Bangladesh; in their deep resentment over the allotment of Sindh's land to bureaucrats, army officials, non-Sindhis (immigrants and Punjabi settlers); and concerning the low level of investments in rural Sindh. A feature common among the nationalists is that in elections none of the parties/groups has ever been returned to the legislatures by the people. Some of those currently among the ranks of the nationalists at one time sat in assemblies on PPP tickets.

Mohajir Qaumi Mahaz (Mohajir National Movement – MQM)   the MQM, a relatively new phenomenon on the political arena, has been changing the complexion of Sindh's urban politics. Established in 1983, its genesis had begun with the formation of the All Pakistan Mohajir Students Organization (APMSO) at Karachi University in 1979. The leadership of the MQM remains the same as that of APMSO. Women activists of APMSO, like Kishwar Zohra and Zareen Majid, are office-bearers in the MQM.

The MQM's propagation of the philosophy of *mohajir* nationalism, combined with its young and middle-class leadership, has caught the imagination of the urban *mohajirs* – the second and third generation of north and central Indian immigrants. Its most significant achievement has been to give a sense of identity to the *mohajir* community – and above all a sense of security.

A highly organized and mobilized party with an estimated 10,000–12,000 trained cadres (*Frontier Post*, 24 July 1992), it has virtually wiped out other political parties (like Jamaat-e-Islami, Jamiat-Ulema-Pakistan) which had traditionally held influence in Sindh's urban centres. As a champion of the middle class it has a youthful leadership from the urban lower middle and middle classes. It swept the local elections in 1987 and later won almost all of Sindh's urban seats in the National and Provincial Assemblies (in both the 1988 and the 1990 elections). The party is highly centralized: its leader, Altaf Hussain, has become a cult figure, treated as a legend in his lifetime. Given the title of *Quaid-e-Tehrik* (leader of the movement) and called *pir sahib* (spiritual leader), Altaf Hussain has an image that has been aggressively built up, as 'blind faith' in the leader is considered a fundamental principle of the MQM (Farooq, 1990, p. 10). The party has a strict hierarchical system and a strong code of secrecy. It is organized pyramidically into a 12-member *markaz* (central committee), four zones with sectors and units under them. Elected representatives do not necessarily have control over the organizational levels, which have a system of vertical reporting and a top-down line of command. Meetings are held regularly at all levels, with those in charge of zones being the most powerful, reporting directly to the *markaz*. Before being allowed to join, would-be members are observed, tested and then recommended by the local unit and approved by the *markaz* before taking an oath of allegiance and loyalty to the party.

MQM activists, armed with sophisticated weapons, have waged violent running battles for the control of Karachi and Hyderabad, the two largest cities of Sindh where the majority of the *mohajirs* are settled. Their domination of the cities has been expressed through calls for general strikes, when no one in the cities can move (*Newsline*, February 1990, pp. i–iii); openly taking the law into their own hands; intimidating the press; attacking dissenters;[7] and harassing even charitable organizations. At the same time, the MQM has also organized free bazaars for the needy, carried out 'clean the city' campaigns, provided employment for its activists and supporters in municipal corporations, etc. Consequently, the MQM has often been accused of running a 'state within a state'. The June 1992 exposure of its torture cells through an army operation, combined with allegations from MQM dissidents, indicate the extent of its violent tendencies (*The Herald*, July 1992, pp. 28–29).[8]

The MQM'S original intention was to build bridges between rural and urban Sindh and rally together against the domination of the Punjabis. Therefore, not surprisingly, its early show of force was vis-a-vis the Pathan settlers of Karachi who control the city's transport and squatter settlements. The carnage experienced in Karachi during that period was unprecedented, though worse was to follow over the years. At the time, the MQM was making overtures towards the extreme Sindhi nationalists (JSM). This was followed, in 1988, by a coalition with the PPP, when the latter came to power at the Centre and in Sindh. An uneasy alliance, it was short-lived, ending when the MQM switched sides dramatically in September 1989 by signing an agreement with the PPP's opposition coalition, the IJI (Saleem, 1990, pp. 219–227).

Battle-lines were drawn up after this, with the MQM and the PPP locked in armed confrontation, fought primarily through their students' wings. The level of brutality employed on both sides was indeed gruesome, peaking in February 1990 with killings with sophisticated weapons, kidnappings and torture, burning to death and the use of electric drills (*Newsline*, February 1990, pp. i–viii). Finally, the army chief of Karachi city intervened, arranging and supervising exchange of prisoners.

Subsequently, the divide between Sindhis and *mohajirs* grew wider, with all shades of Sindhi nationalists arrayed on one side against the MQM. This culminated in the bloody clash in Hyderabad's *mohajir* neighbourhood in May 1990. In this PPP government operation for arresting 'terrorists', about 60 people were killed, including 12 women (*The Herald*, June 1990, p. 33).

The major demand of the MQM is the recognition of *mohajirs* as a separate nationality. It has agitated for greater employment of *mohajirs* in government services and police. Other demands include just enforcement of the quota system (under the current system, the percentage share of rural and urban areas in government jobs is specified); repatriation of Biharis stranded in Bangladesh (since 1971, refusing BD nationality); provision of transport and civic amenities; and reopening of the border with India in Sindh.

Since late 1991, following the army's cleanup operation, the MQM has been rocked by an internal split. The dissident faction, calling itself MQM-Haqiqi (the real ones), has accused the MQM-Altaf of striving for a separate homeland for *mohajirs* as its ultimate objective (*Frontier Post*, 14 January 1992). An army spokesman reportedly confirmed the allegation: 'there were confirmed intelligence reports that the MQM leadership was actively pursuing the idea of carving out some areas of Sindh for the purpose of declaring them as *Jinnahpur* or *Urdu Desh*' (*Nation*, 18 July 1992). The MQM-Altaf denied these charges, saying that there was at one time a proposal to create a separate administrative unit comprising Karachi and Hyderabad. This was being considered as a means of dealing with the problem of unemployment, as the only way to generate enough jobs. But at no stage, they insisted, did anyone call for a separate homeland (*The Friday Times*, 23–29 July 1992, p. 7). However, in February 1994, the Karachi-based leaders of MQM-Altaf publicly voiced the demand for the division of Sindh and a separate *mohajir* province. This was denied yet again by Altaf Hussain, who leads and directs the party from London, where he has been living since the split.

The MQM is a curious phenomenon. On the one hand, as a highly centralized, secretive, disciplined and feared party, it is seen as having fascist overtones; but on the other, it also has a huge vote bank and vast grass-roots support.

10.4   Ethnic Politics and Women's Participation

Historically, women's participation in politics in Sindh has been limited to elite women, and that too in urban areas. The 1980s witnessed a marked change. For the first time in Pakistan's history, there came large-scale grass-roots mobilization of women: rural women have been organized, as have women of the urban petty bourgeoisie, traditionally the most cloistered. The formation of Sindhiani Tehrik and the women's wing of the MQM is significant, not only from the perspective of broadening the base of the political process in the country but also from that of the women's movement – still in its formative stages.

10.4.1   Sindhiani Tehrik[9]

Sindhiani Tehrik, probably the largest women's organization in the country, has demonstrated its ability to mobilize several thousand women at a time. It was formally organized and launched in May–June 1982 as a women's organization in Rahuki, District Thatta, though no name was decided for it at the time. Its first public action was taken in September 1983 during the Movement for the Restoration of Democracy against Zia's government.

The beginnings of Sindhiani Tehrik can perhaps be traced to the mid-1970s, when the Sindh Awami Tehrik (later called Pakistan Awami Tehrik, now referred to as Awami Tehrik) was being formed by left-wing intellectuals and political elements of Sindh. The leadership of the Awami Tehrik – in particular Rasul Bux Paleejo and Fazil Rahu – felt the need to make women aware of the issues of the day: the language issue, medium of instruction, auction of land to non-Sindhis, the question of Sindhi nationalism. They recognized the importance of involving women, partly to create an understanding of their own work among women and partly to expand their political work. Initially women belonging to the families of Awami Tehrik members were approached and mobilized. In 1976, a two-day women's conference was convened in

Hyderabad, to which participants were invited from all parts of Sindh – representing the peasantry as well as urban professionals. During this meeting, apparently men did all the cooking and also took care of the children.[10]

Other such meetings followed, with a cross-section of Sindhi women invited to attend. The objective was to politicize women and to raise their consciousness. Heroines of Sindh were identified and celebrated – like Mai Bakhtawar, who had defied the British. Early attempts were made to form a broad-based organization covering all sections of women, but it soon became clear that urbanized middle class and elite women were not interested in working in the rural areas or doing political work at all. The decision, therefore, was made to concentrate on women of small towns and villages.

The milestone in the emerging Sindhi women's consciousness not only as women but also as Sindhi nationalists was the Sindhi Women's Conference, organized in April 1982 in Karachi. The conference brought together women from all parts and all classes of Sindh. It unequivocally slated the feudal and semi-feudal structures of Sindh as responsible for the oppression and suppression of women. A Sindhi Women's Association was formed, which, however, never got off the ground. The women who later formed the Sindhiani Tehrik had participated in the conference and had been strongly influenced by it. Only a month or so later, they called the meeting in Rahuki where the future Sindhiani Tehrik was launched; Shahnaz Rahu was made the first convenor/President of the Organization.

Sindhiani Tehrik has a well-defined organizational structure. At the top is a 22-member Central Committee, with district units below and *tehsil* (the lowest unit in the administrative structure) units at the lowest level. Members of the Central Committee are elected by district units for a period of two years. The Committee is the decision-making body; it usually meets every two months but emergency meetings may be convened whenever necessary. Although initiated into the political field by men, Sindhiani Tehrik operates independently and makes its own decisions regarding action. It, however, sees itself as closely linked with the Awami Tehrik (referred to as the 'mother party' by Zahida Sheikh, secretary-general of Sindhiani Tehrik), and guided by its philosophy and ideology. The parent body's influence on Sindhiani Tehrik cannot be denied – it actively participated in Rasul Bux Paleejo's election campaign in 1988, enthusiastically celebrates his birthdays and unquestioningly accepts him as 'our leader'.

Membership is made up of professional women and students, peasant women, school teachers and educated housewives. Sindhiani Tehrik claims to have a hard core of 4,000 activists, and a total membership of 50,000 women. In the leadership are women belonging to the middle-level landowning class and the petty bourgoisie of Sindh. Peasant women have over the years moved into leadership positions. Initially, those in the vanguard were close relatives of the former Awami Tehrik's leaders. Since the elections in 1992, however, the promotion of more grass-roots leadership is in evidence; indeed, the current president, Mumtaz Nizamani, comes from the peasantry. Zahida Sheikh, the general-secretary, is in her early 20s, and was inducted into Tehrik's politics through the student wing of the Awami Tehrik. Some of the women, like Akhtar Baloch, had been politicized during the anti-Ayub agitation and were imprisoned for participating in student protests in 1970. Ghulam Fatima, another active member, was a member of the District Council of Thatta. She was also the first woman to court arrest after the harsh repression that followed the MRD movement in 1983.

The major objective of the Sindhiani Tehrik has been politicization of women. Emphasis has been on interaction between women from small towns and their village counterparts. There is constant travel undertaken by the town-based members to the rural areas where meetings are held and study circles conducted. Discussion topics include the ills of feudalism, the roots of oppression, the fundamentals of socialism and, in the earlier years, literature about Russian and Chinese women. Women's literacy is encouraged. Islamization and its implications for women, discriminatory legislation, the role of the *maulvi* (priest) and the use of religion to subordinate women have also been taken up as part of the consciousness-raising sessions on women's issues. While the Sindhiani Tehrik recognizes men's oppression of women, it views this as part of a larger unjust and oppressive system where women's rights cannot be isolated from men and society at large.

Besides meetings and discussions, the Sindhiani Tehrik produces pamphlets, leaflets and newsletters to disseminate its views. Interestingly, it has used religious meetings as occasions to reach out to women. Sindhiani feels that religion is very deep-rooted in their culture; since it has been used to subjugate women for so long, it will take a considerable time to change attitudes instilled through religion.

While most of Sindhiani's reported public action (rallies, demonstrations, protest meetings) has concerned political issues, like the Kalabagh Dam and the repatriation of Biharis from Bangladesh, it has demonstrated for women's issues too. Its student wing, for instance, has agitated for more places for women in medical colleges. Similarly, it has responded enthusiastically when invited to attend WAF meetings in Karachi (50 members came for the International Women's Day meeting in 1986). In Lahore, its representatives participated in the reflective workshop organized by WAF in November 1991 to assess its ten years of activism.

The outreach of the Sindhiani Tehrik spreads to the strongholds of Awami Tehrik. According to its activists, the impact of Sindhiani Tehrik has been quite marked. Considering that Sindh is strongly feudal and the movement of women outside the home in landed families is actually forbidden, Sindhiani Tehrik has brought about a visible change. Women's mobility has increased: even teenage members of the organization travel, in twos and threes, from one city to another with ease. Zahida Sheikh has confirmed this in an interview, 'I was not allowed by my parents to attend Sindhiani Tehrik's first convention. During this whole period I had to fight and resist the orders of my parents, but later my mother . . . realized my position. Now if I go to my village after three months they will not enquire about my absence' (*Viewpoint*, 30 January 1992, pp. 10–11).

The *burqa* (the long, loose dress worn as veil and covering over face and clothes) is vanishing among the educated, and the self-confidence of Tehrik members has grown proportionately with their activism. On the occasion of Paleejo's birthday in 1989, Sindhiani Tehrik members marched with Kalashnikovs, marking perhaps a new level of militancy. In January 1992, a large contingent of Sindhiani Tehrik participated in the 700 km long march from Sukkur to Karachi organized by the Awami Tehrik, to raise awareness in Sindh.

The Sindhiani Tehrik has also received its share of criticism. The organization, and particularly its activists, has been blamed for arousing rebelliousness among women. They have actively campaigned against early marriages, polygamy (even though Paleejo has been married four times) and childhood betrothals – all norms that have formed part of Sindh's social tradition – and have advocated education for women and women's consent in marriages. Sindhiani Tehrik admits that it is difficult to fight 'these social evils so deeply rooted in our society', but nevertheless tries to 'cultivate the minds of society' to rid it of the feudal practices (*Viewpoint*, 30 January 1992, vol. 17, no. 25, p. 11).

10.4.2   The MQM Women's Wing

Women's role in the MQM started with the inception of the organization. Zareen Majid, Vice-Chairperson of the MQM and member of its Central Committee, was, like other founders of the MQM, an APMSO activist. The women's wing was set up in 1986 following bloody clashes between MQM workers and residents of a Pathan squatter settlement on the outskirts of Karachi. This was perhaps the first experience of *mohajir* women's active participation in politics. The wing established its real effectiveness during the local election of 1987 when 'the traditionally conservative women of the *mohajir* community joined in whole-heartedly with their men in the local election campaign' (Azam Ali, 1988). They also voted in the election which the MQM won hands down, both in Karachi and in Hyderabad. The success was attributed in equal measure to women's efforts: the MQM leader, Altaf Hussain, was in prison at the time, and women campaigned behind the scenes for his release, sending thousands of letters and telegrams to the Chief Minister of Sindh.

The first public demonstration of the women's wing was the mammoth rally held on 15 July 1988. Attended by from 'anywhere between 50,000 to 100,000', the rally was seen as the MQM's show of strength vis-a-vis the Jamaat-e-Islami and Jamiat-Ulema-Islam, its political rivals in urban Sindh. But it was equally a display of organizational capacity of the women's wing, who had managed to bring together women from all over Karachi. Armed female guards controlled the crowds, showing their readiness to join the 'Kalashnikov politics' of Sindh (Abbas, 1988).

A second expression of women's militancy was in the Pucca Qila (Strong Fort) incident of 27 May 1990 in Hyderabad. Whether members of the women's wing were involved in the action is not certain, but women came out, carrying copies of the Quran in their hands, leading a procession of men and children to protest against the disconnection of the water supply in the area. Hyderabad had been under curfew since 15 May (312 hours, with only 20 hours of breaks). When this procession moved towards the Qila (fort) gates it was met by 'a waiting line of policemen' that ordered the women to halt and then opened fire at point-blank range when the women refused to stop. At least 12 women lost their lives that day (*The Herald*, June 1990, p. 34).

Since then, in moments of crisis, women activists and those belonging to the families of MQM cadre have come out in public protest. As the leadership moved underground following the army operation, women have managed the party's secretariat; about 150 are said to be underground and wanted by the army (*Newsline*, May 1994, p. 30).

The fact that the work of MQM spreads across *mohajir* neighbourhoods facilitates women's participation. Often, the involvement of the men of the family in politics has encouraged women to join. As with most other political groupings in the country, women in the MQM are related to the men in it. As a woman at the 1988 rally said: 'When our children are in the firing line, and our men are so committed, how can we stay away?' But other women, like social workers in *mohajir* neighbourhoods, have also been drawn in; and some have joined despite family opposition (Azam Ali, 1988).

The women's wing does not have a separate programme but functions as an adjunct of the main MQM, playing what appears to be a secondary role. Kishwar Zohra, head of the women's wing in Karachi, said that *mohajir* women's problems are like those of MQM men. According to Zohra:

There is a higher percentage of educated girls among the *mohajirs* community than in any other. We have many MBBS [qualified doctors] and other graduates who are jobless; many of our girls face the same difficulties as the boys in getting admissions.

The same sense of cultural identity and the feeling of being discriminated against which have aroused the passions of the *mohajir* men, also agitate the women. Older members look back on these 40 years in Pakistan and say that their dreams and hopes have not been fulfilled. (Azam Ali, 1988)

Zareen Majid, Vice-Chairperson and member of the Central Committee, was the only visible female office-bearer of the party until the split, when she was sidelined. It is not clear whether women now form part of the high command or not. Two women were elected from the MQM platform in the 1993 elections, one to the Provincial Assembly and one to the Senate, but neither seems to have a leadership position. In 1990–91 there used to be about 100 separate women's units spread across the neighbourhoods of Karachi. Each unit had three office-bearers, who held regular meetings with the MQM hierarchy where policy and directives were passed on for action. Women members of MQM would take an oath of loyalty in the same manner as male members. Instructions were also given to women regarding the mode of dress (the head must be covered with *dopatta* – head scarf), attitude (patience, willingness to listen) and behaviour (politeness) when contacting other women in their neighbourhoods. The state of the party structure, ever since the army's action and the breaking of the party, has not been certain.

The women's wing has always been highly opaque. Names of women leaders, with the exception of Zareen Majid and Kishwar Zohra, are not generally known. The only new ones coming to public knowledge are those of Feroza Begum and Nasreen Jaleel, the two elected women. None of the women is easily accessible for interviews or discussions; in fact, perhaps the only time that Zareen and Kishwar were interviewed by the press was following the July 1988 rally. When this author tried to meet them, she was told first to get permission from Altaf Hussain or Azeem Tariq (President of MQM) – not an easy task, as neither was readily available. Finally, when an informal interview with the latter two was arranged, permission was not forthcoming. The women who protested army action in February 1994 insisted on anonymity while expressing their anger to the press. In fact, when pressed to answer they said, 'you should ask these questions of our men'. And on their views about the solution to MQM's problems, the response was the same: 'That is the concern of our men. You ask them' (*Newsline*, May 1994, pp. 30–31).

As there is very little information available about the women's wing it is difficult to know the extent to which women's issues feature in women's organizational meetings or even in the MQM structure. Issues like dowry and divorce are apparently discussed (Azam Ali, 1988). Unsubstantiated reports say that the MQM influences the lives of families in its areas of control. Marriages, for instance, cannot be arranged without the approval of the local MQM unit, and differences and conflicts are often resolved by it. No more is known about the involvement of women's units in such situations.

Similarly, there is no indication whether women reacted when MQM's sitting legislators took a public oath of allegiance to Altaf Hussain, at a five-star hotel in Lahore, after the split in MQM became public. In all, 11 legislators solemnly proclaimed their loyalty to the *Quaid* (leader) declaring that 'betrayal with the chief of MQM is more shameful than raping our own mother, sister and daughters' (*Frontier Post*, 25 July 1991).

Nor is it known if there was a debate within the MQM on the controversial Shariat Bill which was bulldozed through the National Assembly on 16 May 1991. This bill had been in the pipeline since 1985 and sought to ensure the supremacy of the *Shariah* (Islamic law) in the country. Challenged by women who see within it the seeds of gender discrimination, it has been opposed by progressive women's organizations, the PPP and almost all the Left parties. The MQM had also publicly opposed the Bill, seeing it as an instrument of discrimination against women (WLUML, 1992, p. 3). However, in the Assembly the MQM chose simply to abstain from voting rather than take a position against it.

Interaction of the MQM's women's wing with other women's groups has been negligible. Whereas most women's groups and sections of political parties have in the past decade met either in collaboration or in confrontation, there is no recorded instance of any meeting point with the MQM women's wing.

10.5 Conclusions

From the above review it seems clear that the creation of both the MQM women's wing and Sindhiani Tehrik came in response to the felt need of the two ethnic movements to maximize their impact and generate a groundswell. However, this should also be seen in the context of the larger political environment in Pakistan: repressive military dispensation, proliferation of drugs and arms, the pressures of urbanization, unemployment, shortage of educational institutions and the confronting of the military regime by women. The last factor is significant in that women, under the umbrella of the Women's Action Forum, were the first organized group to challenge martial law orders and ordinances (Mumtaz & Shaheed, 1987, pp. 106–108). Public protests and high-profile activism by women not only placed women's issues on the political agenda of the country, but also established higher thresholds of acceptability for women's participation in politics.

It is therefore not mere coincidence that Sindhi women had not been able to organize successfully prior to 1982. The necessary ingredients – besides the intensity of ethnic consciousness – like women's preparedness to participate, state oppression and an environment ready for accepting women's mobilization, seem to have come together at that point and time. These are factors which in all likelihood have also affected women's induction in the MQM. Yet, not unlike the experiences of various movements across the globe (Isaksson, 1988, p. 5) the roles of women in Sindh's conflict are conceived and determined by men, as supportive and subordinate – for creating understanding and for galvanizing women for 'the cause'. The parent parties initiated the creation of women's sections and have subsequently provided direction.

Sindhiani Tehrik and MQM are quite distinct from each other despite common motivations for their inception. What distinguishes them most is their mode of working. The Sindhiani Tehrik is a relatively open organization, with its activities reported in the press, its leadership accessible to members, press and individuals. It keeps contact with other women's organizations and is responsive to invitations from, for example, WAF and non-Sindhi women's rights groups. Working within the parent party's political and strategic framework, it operates in a relatively autonomous manner. Its internal workings, like holding of elections, are decided by its Central Committee. The MQM women's wing, by contrast, is shrouded in secrecy. It operates, much like the MQM, in a closed structure and has not expressed any measure of autonomy. Neither is it known to have contacts with non-MQM women's groups, and it has remained almost completely inaccessible to outsiders.

The two nevertheless share some characteristics. For instance, their class backgrounds; their leadership comprising first-generation educated women; their energy, commitment and militancy, as evident in public demonstrations and rallies; their ability to handle weapons.

With reference to women's involvement in movements, several questions are often raised. Does the participation of women help to reduce the levels of violence? Or find creative ways of conflict resolution? When the crisis ends, are the women relegated to their former position? Or will the nature of their involvement lead to a longer lasting impact? Such questions are often asked without taking into consideration existing social structures. When women play a subordinate role in society and their 'terms' of participation in move-ments/struggles are determined by men (as in Sindhiani and the MQM women's wing), then they are not likely to influence the directions that the movements take, nor the level of violence within these, nor mechanisms for conflict resolution. Under such circumstances the role of women is basically to serve as 'propagandists', and their participation is based on 'order and authority' – with none or at best token inclusion in decision-making bodies. Experiences like the Algerian Revolution have shown that the fact of partic-ipation in struggles, even socialist ones, does not necessarily pave the way for women's emancipation. Helie-Lucas writes of how 'obedience, morality and conformity . . . necessary conditions to be part of the revolution' set the 'roots for a tightly controlled society' (Helie-Lucas, 1988, pp. 178–179). And how policies regarding women in post-liberation Algeria were defined or re-defined in accordance with the needs of the ruling class (p. 183). It is not surprising that women return to the hearth, so to speak, after such struggles – be it on the South Asian subcontinent (Jayawardena, 1986) or in Chile (Bunster, 1988). This could be a possibility also in the case of the two groups under discussion here, except that the existence of an autonomous national women's organization (WAF) does create a different dimension (Mies, 1986),[11] offering the option for already mobilized women to continue the struggle for women's rights.

Even though it has campaigned against the institutionalized oppression of women, Sindhiani Tehrik, for instance, does not believe that its present strug-gle will necessarily change the situation of women: 'these social evils [are] so deeply rooted in our society' – as noted in section 10.4.1 above. The implica-tion is thus that the struggle for women's rights and liberation will have to continue, even after the more immediate objectives of their ethnic/nationalist struggle are achieved.

The Sindh conflict is an ongoing one, so the nature of each experience will determine the future course of action for the women involved. Sindhiani members, for instance, seem to be creating a space for themselves in society through enhanced mobility, building up a relatively independent organiza-tion, raising women's issues, which may outlast the politics of ethnicity. In addition, as involvement in movements does form a collective memory, these can have long-term implications for women. Such experiences become the links in the 'continuous process aligning forces and defining movements that foreshadows the events of today' (Mumtaz & Shaheed, 1987, p. vii). This contention is substantiated by the fact that many of the activists in Pakistan's contemporary movement for women's rights (WAF) are third- or fourth-gen-eration members of families who have been active on the women's front and the nationalist movement over the past several decades (Mumtaz & Shaheed, 1987, p. viii).

On the other hand, with the current heightened levels of politics of ethnicity in Sindh, where no immediate signs of resolution are visible, the major priorities for women will continue to be those of the respective ethnicities. And they will still be set by the leadership rather than by the women themselves.

Notes

1. Out of 1.48 million acres made cultivable by the Ghulam Mohammad Barrage, 0.87 million acres were allotted to defence personnel, bureaucrats and those displaced by the building of the new capital in Islamabad, by Tarbela Dam and Mangla Dam. Out of 0.28 million acres made cultivable by Sukkur Barrage, 0.13 million acres were given to army personnel.

2. Reserved Quotas for Civil Service Selection (*The Herald*, January 1988, p. 86): Merit: 10%; Punjab: 50%; Sindh (urban): 7.6%; Sindh (rural): 11.4%; NWPF: 11.5%; Baluchistan: 3.5%; Federal Administrative Tribal Area (FATA): 4%; Azad Jammu & Kashmir (AJK): 2%. (The quota system was introduced by Bhutto in 1973 for a period of ten years. It was extended by General Zia-ul-Haq for another ten in 1983.)

3. Political and ethnic violence jacked up the levels of all crimes in the province. In Karachi, for instance, crime soared between 1985 and 1990. From two cases of kidnapping for ransom in the city in 1985, there were 91 cases in 1990; 220 murders in 1985, as against 585 murders in 1990; 1,553 motor vehicle thefts in 1985, but 8,091 in 1990 (*Newsline*, October 1991, p. 30.)

4. The fronts in Jiye Sindh Supreme Council are:

- Jiye Sindh Mahaz (Political Front)
- Jiye Sindh Students' Federation
- Jiye Sindh Girls' Front
- Latif Students' Sangat (Children's Organization)
- Nari Tehreek (Women's Front)
- Poreet Sangat (Labour Front)

All the above are directly responsible to the Supreme Council which is composed of two members from each front (Rana, 1990, p. 13).

5. The student organizations operating in Sindh are (Rana, 1990, p. 14):

- Jiye Sindh Students' Federation (JSSF)
- Sindh People Students' Federation (PPP)
- Sindhi Shagird Tehrik (Awami Tehrik)
- New Sindhi Students' Organization (Punjabi)
- All Pakistan Mohajir Students' Organization (APMSO)
- Sindh Watan Dost Student Federation (Communist Party)
- Pakhtoon Students' Federation (Punjabi-Pakhtoon Ittehad)
- Baloch Students' Organization (Independent)
- Democratic Students' Federation (Communist Party)
- Sindh National Students' Federation (Sindh National Front)
- Muslim Students' Federation (Muslim League)

6. The dam, a major hydro-electric and irrigation project, would drastically reduce the waters of the Indus River, affecting agriculture, the Indus delta and aquatic life in the coastal areas of Sindh.

7. In a press conference addressed jointly by 37 MQM members of national and provincial assemblies, a number of members said categorically, 'we will not hesitate even for a moment to destroy the houses of dissident leaders'; and that 'the minimum punishment for dissidents is death' (*Frontier Post*, 25 July 1991).

8. The split in the MQM followed the decision to expand the base of the party to attract

supporters from other provinces with the promise to 'establish a system . . . beneficial for the 98% poor and middle class masses', and to eliminate 'this obsolete and out-dated feudal system from its very roots' (*Frontier Post*, 21 May 1991). It therefore decided to change its name to Mutahedda Qaumi Mahaz (Joint National Front) from Mohajir Qaumi Mahaz. This decision, made public in May 1991, was opposed by a section of the party. Three of the MQM provincial assembly members publicly challenged the new policy which they saw as undermining the *mohajir* cause. Infighting ensued, ending in the killing of eight persons in June 1991 (*Frontier Post*, 18 June 1991).

9. This section draws heavily from a paper by the author entitled: *Khawateen Mahaz-e-Amal (Women's Action Forum) and Sindhiani Tehrik: Two Responses to Political Development in Pakistan*. It was written for the 1989 Annual Meeting of the Association for Asian studies, panel on *Women and Pakistan's Development*, Washington 17–19 March, and was subsequently (1992) updated for publication in *South Asia Bulletin*. The information derives largely from interviews conducted with various members of Sindhiani Tehrik, supplemented with printed material (pamphlets, newspaper reports, published interviews, etc.).

10. There are conflicting reports on this. While the women interviewed by the author said that men cooked and looked after the children, others are reported as saying that men arranged only the food and did not help with the children.

11. Maria Mies, discussing the issues of 'National Liberation and Women's Liberation' (Mies, 1986, pp. 175–199) considers the formation of national women's organizations as a touchstone for women's liberation (p. 195). The critical factor, she says, is whether they were formed before a struggle or after (p. 196). Here the case of Pakistan is somewhat different, in that the national women's organization is not part of the ethnic struggle. The movement for women's rights is being waged at a parallel level, independent of ethnic issues.

References

Abbas, Mazhar, 1988. 'The New Vanguard', *The Star* (Karachi), 28 July.

Abbas, Zafar, 1989. 'Personality Interview: G.M. Syed', *The Herald* (Karachi), August, pp. 169–183.

Ahmad, Eqbal, 1992. 'The Challenge in Sindh', *Dawn* (Karachi), 21 June.

Azam Ali, Ameneh, 1988. 'A *Mohalla* [Neighborhood] Movement', *The Star* (Karachi), 28 July.

Bunster, Ximena, 1988. 'The Mobilization and Demobilization of Women in Militarized Chile', pp. 210–222 in Isaksson, 1988.

David, Kumar, & Santasilan Kadirgamar, 1989. *Ethnicity: Identity, Conflict and Crisis*. Hong Kong: ARENA Press.

Farooq, Imran, 1990. *Nazm-o-Zabt Key Taqqazay* (The Compulsions of Discipline). Karachi: MQM.

The Friday Times (Lahore), 23–29 July 1992.

Frontier Post (Lahore), various dates, 24 July 1992.

Gardezi, Fauzia, 1990. 'Islam, Feminism and the Women's Movement in Pakistan: 1981–1991', *South Asia Bulletin*, vol. 10, no. 2, pp. 18–24.

Helie-Lucas, Marie-Aimee, 1988. 'The Role of Women During the Algerian Liberation Struggle and After', pp. 171–189 in Isaksson, 1988.

The Herald (Karachi), various dates.

Isaksson, Eva, ed., 1988. *Women and the Military System*. London: Harvester–Wheatsheaf.

Jayawardena, Kumari, 1986. *Feminism and Nationalism in the Third World*. London: Zed Books.

Mies, Maria, 1986. *Patriarchy and Accumulation on a World Scale*. London: Zed Books.

Mirza, Mahmood, 1986. *Aaj Ka Sindh* [Today's Sindh]. Lahore: Progressive Publishers.

Moghadem, Valentine M., 1992. 'Patriarchy and the Politics of Gender in Modernizing Societies: Iran, Pakistan and Afghanistan', *International Sociology*, vol.7 no.1, March, pp. 35–53

MQM, 1990. *Altaf Hussain Ki Kahani: MQM Kay Quaid Say Taweel Interview* [Altaf Hussain's Story: A Long Interview with the MQM Leader]. Karachi: Mohajir Academy.

Mumtaz, Khawar, 1992. '*Khawateen Mahaz-e-Amal* and *Sindhiani Tehrik: Two Responses to Political Development in Pakistan*', *South Asia Bulletin*, vol. 11, pp. 101–109.

Mumtaz, Khawar & Farida Shaheed, 1987. *Women of Pakistan: Two Steps Forward, One Step Back?* Lahore: Vanguard Books; London: Zed Books.

Nation (Lahore), July 1992.

Newsline (Karachi), various dates.

Rana, Sarwar, 1990. 'Sindh Report'. Unpublished paper.

Rashid, Jamil, 1991. 'Ethnic Conflicts: a Case Study of Karachi', *Contemporary Conflicts*. Karachi: Psychiatric Association of Pakistan, pp. 109–114.

Rouse, Shahnaz, 1986. 'Women's Movement in Pakistan: State, Class, Gender', *South Asia Bulletin*, vol. 6, no. 1, Spring, pp. 30–37.

Saleem, Ahmad, 1990. *Sulagta Hua Sindh* [Burning Sindh]. Lahore: Jang Publishers.

Shaheed, Farida & Khawar Mumtaz, 1990. 'The Rise of the Religious Right and its Impact On Women', *South Asia Bulletin*, vol. 10, no. 2, pp. 9–17.

Viewpoint (Lahore), January 1992.

WLUML, 1992. *Special Bulletin on the Erosion of the Judiciary and Human Rights through Legislation (Pakistan)*. Lahore: WLUML–Shirkat Gah.

11

Strategies for Conflict Resolution: the Case of South Asia

KUMAR RUPESINGHE

11.1 Introduction

In the final stages of the 20th century it is indeed pertinent to discuss future directions in conflicts and their resolution. This is a time when profound changes are taking place in the global order, with specific historical conjunctures and a major political move away from bipolarity towards multipolarity. In the long run, these changes will have far-ranging effects on the politics of South Asia as well.

The first part of this chapter seeks to evaluate the terminology of conflict resolution and its relevance to protracted social conflicts. In the second part, some of these concepts are applied to the violent social conflicts in South Asia. I argue that conflict resolution must be seen within a wider framework of analysis which can take into account the ideological and political conditions under which conflicts unfold.

An understanding of conflicts and their resolution in South Asia is of particular significance in a region which has experienced blood and turmoil in truly epic proportions. More conventional disciplines have not managed to provide adequate explanations or predictions for the seemingly unending spiral of violence that has gripped societies in the region. There have, however, been some important scholarly interventions regarding violence, as well as more specific, comparative investigations into the cultural sources of violence. These efforts have focused on the comparative experience from scholars within the region, interpreting modern violence and identity, and making important generalizations (see Das, 1990).

Let us first examine the concept of 'conflict' as it has been defined within conflict theory itself. 'Conflict' is normally identified as a situation where different actors are pursuing incompatible goals. This assumes that the goals or interests are recognized by the parties to the conflict. Interests so defined are objective – whether a conflict over a pay claim between workers and entrepreneurs, or a matter between tenants and landlords, or family disputes between husbands and wives. A further definition holds that conflicts involve apparently incompatible values, where the task of a third party may be to help the parties to specify their values more explicitly so as to facilitate resolution of the conflict.[1]

There are within these conventional paradigms certain assumptions which are largely unstated but which need to be deconstructed. Otherwise, we will be at a loss as to why conflict theory has not been more useful in resolving the violent conflicts we are witnessing today. Or we will be faced with rhetorical questions as to the endemic violence of Third World societies, their undemocratic character and the futility of attempts by these societies to resolve these conflicts in a more humane and peaceful manner. Let us now try to identify some of the unstated assumptions behind such an approach to conflict resolution.

11.1.1   Rationality and Conflict Resolution

The very large and growing literature on conflicts and conflict resolution has largely been in the form of theoretical reflection coming from the USA and Europe. This approach generally presupposes a domain of 'rationality', where all the parties more or less share certain central values based on rational argument. It is assumed that the problem is to get the parties to the table and that, through negotiations, it will be possible to find a win–win solution agreeable to both sides. In this view, the environment within which these conflicts occur is generally founded on a strong ideological imperative of equality and recognition of the rule of law. The modern division of labour in these societies assumes that members are tied to multiple roles and are attached to a variety of interests which result in conflict. Recognizing this complexity, society develops institutions and mechanisms to resolve conflicts in a specified way. Gradually a culture of negotiations emerges, and a complex network of arbitration and dispute resolution becomes increasingly professionalized. Conflicts take place amongst like-minded actors who speak a common language, which may be English. But this concept goes further to denote a shared universe of meaning. Normally disputes are defined within a fairly developed regime of law – mostly individuals' rights – with its specific historical evolution in the West. In modern states, conflicts tend to be symmetric ones in which the state normally assumes the role of a mediator or sometimes plays the role of a third party. However, in the conflicts confronting South Asia today, we find that the state is involved as a party to the conflict.

11.1.2   Conflict Theory as Applied to Protracted Conflict

How relevant, then, are these approaches to the protracted violent conflicts which we are experiencing in the region? These conflicts are not merely based on interest, but involve many social dimensions which concern identity and security. Conflicts which involve a core sense of identity between or among parties tend to be intractable: the intractability is generated by the very dynamics of the conflict, rather than by a reasoning process of rationality. Conflict resolution here means changing the conditions of intractability. These are not single-issue conflicts, but are better characterized as multiple conflicts which are being waged simultaneously.

11.1.3   Armed Conflicts and Their Characteristics

During the summer of 1994, Wallensteen and Axell identified 32 contemporary armed conflicts, with armed conflict defined as a situation with over 1,000 casualties.[2] The total number of armed conflicts seems to be on the increase, especially if we also consider armed conflicts with less than 1,000 casualties. According to Wallensteen and Axell, the total number of conflicts between 1989 and 1993 was 90, in 61 locations around the world. In 1993 there were still 47 continuing conflicts. However, the incidence of direct violence as measured by casualty rates cannot provide a full picture of the scale of social violence.

Most of the armed conflicts take place in Third World countries. Moreover, in the overwhelming majority, the basic issues in the armed conflicts of 1993 were related to internal matters. On the basis of these observations, several generalizations can be made which we can contrast with the conflict environment based on the paradigm of rationality.

In all cases, the conflicts are between the state and particular social groups: frequently between ethnic groups and the state. Although these are frequently characterized as conflicts between two parties, i.e. the state and a particular ethnic group, there are a range of multiple conflicts involving violence, with many parties and many issues involved. In these it can be difficult to identify so-called 'actors', as the more articulate ones with command over resources and communications have a better chance to make known their claims. The Tamils and the Sikhs, for instance, have been able to advertise their conflicts more successfully than many others. Generally the conflicts involve non-negotiable and intangible demands such as identity, security and social justice. In some cases, the demand is for a separate state or for some form of regional autonomy.

The environments within which conflicts occur are highly non-egalitarian, without any all-pervasive rationalist discourse. Often it is a multi-lingual and multi-cultural environment where different meanings are attached to the discourse and where communication may be not only between the parties but also within the parties themselves. The objective seems to be to consolidate political power within one's own community. The discourse of conflict then is fragmented and disjointed, multifaceted. The state, ill-equipped to deal with the violent claims made upon it, tries to resolve the crisis through military means. At no point are non-military means excluded from the range of available options. There is also no solid tradition of governance and little respect for law and order.

Here I have of course exaggerated the differences, if only to delineate the specificity of the conflict environment that we are dealing with, whether it is within the so-called Third World, the former Soviet Union or – not least – Eastern Europe. What is obvious is that protracted social conflicts today are about power and about politics.

11.1.4   A Reconceptualization of Conflict

Earlier on I referred to definitions of conflict which have emerged largely from the discourse of rationality. Now I would suggest that, in the most general terms, we should see conflicts as *collisions between projects*. Projects are sequences of actions directed towards a goal; conflicts occur when the projects of different actors start impinging on each other. We may take missions as an example. Missions are projects of the largest historical scale. Their space is the world, their time measured in millennia. Among the world religions, two stand out as missionary creeds: Christianity and Islam. Or take ideologies: if there is anything distinctive of this century it is the great battle of ideologies – whether they be capitalism, communism, fascism, democracy or the idea of progress. Such missions and ideologies tend to be universal and look to the future, whereas ethnicity, caste and tribalism all tend to be inclusive and restricted, looking more to the past.

It is a truism which bears repeating that all conflicts have a beginning and an end. We can speak of conflict processes, conflict formations, conflict escalation, and conflict termination and renewal.

11.1.5   Modern Conflicts

In retrospect, it is easier to demarcate the termination of a conflict rather than its origins. Although historians may differ over details, I think that there is a fair amount of consensus regarding the great historical conflicts of the past. Only with the French Revolution do we see the emergence of social history, history being made from below, and the beginnings of the great social revolutions. With the French Revolution surely came a redefinition of sovereignty. The great ideological debates over freedom, equality, democracy and much of what is happening today may be traced to this period in history.

The great conflicts of the 20th century are too well known to require elaboration here. The main historical and ideological movements and projects have been the capitalist project and the socialist project, the right to self-determination and nation-building, and the democratic project. What is significant in such projects is their *universal* character. What is distinctive about these visions – whether they are projects over *self-determination*, over *capitalism*, or *democracy* – is their common acceptance of *modernization*, whether it is specific to nation-building, socialism or capitalism. They all subscribe to the building of a future. Let us now turn to some of the general aspects of modern conflicts.

11.1.6   The Capitalist Project

The most significant historical development has been the capitalist project, a truly universal project which knows no boundaries. Within this project there have of course been major differences as to the content of capitalism, leading to two world wars. Today, however, the debate is not over the significance of capitalism as a universal project but an analysis of the various transitions within the trajectory of capitalism. The transition to capitalism is difficult, and understanding the different trajectories and phases within this transition will require analytical rigour. There are many societies which may never successfully achieve the transition – societies sometimes designated as the Fourth

World. Transitional situations are fluid, and conflicts are the stuff through which the content of this evolution is determined. The scale of capitalist transformation may extend for several decades into the future. With the universalization of capitalism, the latest historical project has assumed hegemony – there are no alternatives except a new phase which may be characterized as 'post-modernity'. With the growth of towns and cities, the development of communications and the 'information society', we have a radical rupture with tradition. Post-modernity refers to a particular phase in the development of capitalism where automation and the conflation of time and space lead to greater diversity and atomization. Especially relevant to us here is that decisive conflicts may not be over territory but over other forms of control and regulation.

The most significant alternative to the capitalist project was the socialist one. The mammoth Soviet Union – with its hegemony over the states of Eastern Europe – acted as a centre for the alternative, so-called non-capitalist road to socialism. This project had its internal dissidents and splits, and is now in the process of disintegration. Basically, this alternative project has collapsed.

11.1.7 The Right to Self-Determination

The project of self-determination began with the French Revolution. Its objective was to redefine sovereignty and to insist that sovereignty did not radiate from God through the monarch, but rested with the people. The principle of self-determination has been widely accepted since the appearance of democratic ideologies in the American and French Revolutions of the late 18th century. In the American Declaration of Independence, it was treated as axiomatic that 'the Laws of Nature and of Natures' God' entitle 'one People to dissolve the Political Bands which have connected them with another, and to assume among the Powers of the Earth the separate and equal Station' as the fledgling United States then claimed. A century and a half later, in 1918, President Woodrow Wilson persuaded the Allies to adopt self-determination as a war aim. In his Fourteen Points, he called for the independence of Poland, the restoration of Belgian sovereignty, the adjustment of Italy's frontiers along recognizable lines of nationality, the re-establishment of the Balkan states on historical lines of allegiance and nationality, the granting of free opportunities of autonomous development for the peoples of Austria-Hungary, and the assurance of the same right to peoples under Turkish rule.[3] In an address to the US Congress, he generalized from these particular cases:

> National aspirations must be respected, peoples may now be dominated and governed only by their consent. Self-determination is not a mere phrase. It is an imperative principle of action, which statesmen henceforth ignore at their peril.[4]

The right to self-determination remains one of the most intractable and difficult issues to be addressed by the international community. Many legal formulas have been explored to define the existence of the right to self-determination, to define who constitutes a people, and who has a right to a separate existence. The subject has been the basis for contention and for many a war.

Officially, the United Nations' decolonization project has been wound down with the independence of Namibia. However, there are some notable exceptions which claim special status, such as the Palestinian question. Recently the Kurdish question has once again highlighted the fate of non-state peoples and the inadequacy of ad hoc international efforts at protection.

Colonial and alien domination was treated as a phenomenon that applied only where the dominator was European – with two exceptions, South Africa and Palestine. A distinction is drawn in practice between so-called 'salt sea' imperialism, where the dominating and dominated are separated by hundreds of miles, and local imperialism, where the two peoples are immediate neighbours. Until very recently it has been assumed that peoples locked together within a state must remain so linked indefinitely. This means that many cases of 'internal colonialism' do not come under the purview of any international body. Now, after the decolonization process, new groups are claiming sovereignty and the right to self-determination. It is here that there is a proper concern as to whether the project should be extended ad infinitum. The project on self-determination is still not complete, and, if adequate protection is not found for national minorities, this is likely to be a source of violent conflicts in the future. It is my contention that adequate safeguards for the protection of minorities are *not* provided, and that the international system is still bound to a concept of state sovereignty which paralyses it in action – particularly when it comes to ensuring the adequate protection of minorities. Unless these questions of adequate guarantees of security, reasonable autonomy and linguistic rights can be resolved, we will experience large-scale violent conflicts within the next few years. Whether in the former Soviet Union, the former Yugoslavia, in certain parts of Asia, or in Africa, the future looks bleak.

11.2   The Role of the State

The modernization project has been accompanied by a highly centralized and standardized bureaucratic system whose apotheosis has been the development and articulation of a centralized state. The evolution of this centralized state was the project which captured the imagination as the best vehicle for the evolution of human civilization. It became the vehicle upon which violence was mediated between itself and the people, through the evolution of a technocratic bureaucratic structure which has taken upon itself a monopoly on violence. The evolution of the state and the process of standardization meant that cultures and languages were either absorbed, eliminated or incorporated into the modern project, and this continues. The state-building project is not yet completed, and there are many new nations which are demanding state sovereignty. The project 'one nation, one state' continues to evoke passions and mobilize people.

What is new is that the process of centralization and state-building is increasingly challenged by a variety of social and ethnic movements. The consolidation of state power in the future is problematic, for several reasons:

- The concept of sovereignty is being gradually eroded.
- The unitary state as a powerful centralizing agency is under challenge from subnational forces.
- Violence is no longer the monopoly of the state alone; various transnational and subnational forces are able to arm and equip armies and deliver lethal weapons.

11.2.1   The Concept of Sovereignty

The modern state system has European origins. From a small number of states, it has today expanded to a proliferation of states – which itself constitutes a major global project of universal dimensions. The state-building project has assumed new vigour after the Cold War, with a number of new states emerging. However, under conditions of modernity there has also been an erosion of the concept of sovereignty, such as non-interference in internal affairs, and the prerogatives of the state have been challenged by many institutions. The metaphors of the 'global village' and modern communications have helped to serve this purpose. Furthermore, international institutions, which began as complementary to state-building projects, have assumed an autonomy of their own, sometimes imposing their will on individual states. In the domain of human rights and humanitarian intervention, norms are being developed and states are scrutinized for their human rights performance.

11.2.2   The Unitary State

The process of state-building has been characterized by strong centralization and bureaucratic management. Often, unitary state structures have been controlled by hegemonic elites who may marginalize those on the periphery and other identities. This process of unitary state-building has often imposed the principle of one language, one nation.

State formations are now in different phases of evolution. Some formations have achieved a high degree of integration, as we see with today's European Union, where border controls for those within the EU have been abolished. Most states, however, are in different phases of evolution. Often states are dominated not only by bureaucratic centralization but by hegemonic elites with wide patron–client networks which exclude other nationalities. Some of these states may evolve into truly multi-ethnic societies, although the idea of the 'melting pot' as a paradigm for social integration may not be relevant to all segmented and deeply divided societies. The uneven development of state formations means that there are highly developed states (often called the democratic zone), and states in formation, and states yet to be born. Reform of the international system means that we shall have to recognize this fact. Whilst some developed states may transfer sovereignty to higher bodies, others may cling to a more narrow definition of sovereignty.

Most of the emerging conflicts concern the nature of the state and its formation. Whether the conflicts are over the devolution of power, federalism, governance or how resources are distributed, they generally concern the way in which the state manages its business. Several states are themselves a result of violence and bloodshed. Some states are hegemonic states, based on communal/ethnic, or religious loyalties, where patterns of recruitment are based on ethnic affiliations. Some states can be called defective states, in that they continue to preside over their own retardation. But in general all states are confronted with the same basic challenges. The most significant one is the requirement for modernizing their economies within a globalization process which is accelerating and frenetic. One internal threat comes from the military, and another from ethnic and religious fundamentalist forces. These two movements therefore constitute the twin challenges to democratic development. A further observation is that, in dealing with these issues, the state has become an agent of arbitrary violence, perpetuating violence and militarism as a way of resolving conflicts.

There is yet another significant reason why conflicts of the kind we mention are becoming increasingly unmanageable: the proliferation of weapons and the diffusion of weapons technology. Today, new actors are determining the direction of conflicts. There is a growth in transnational networks that trade especially in small weapons and are often linked to the drug trade.

11.2.3   The Democratic Project

The democratic project evolved concurrently with capitalism. Whereas the early phase of democracy was specific to Europe, it soon expanded to North America and is – in its current phase – expanding as a global project. There are different variants within the concept, but as a metaphor it emphasizes the sovereignty of the people, equality and freedom expressed through periodic elections, and respect for constitutions. There are also different explicit variants, ranging from the most ideological one – that of liberalism identified with the free market – while other variants include forms of social democracy. Paradoxically, democratization also creates space for ethnic revivalism and for religious fundamentalist movements. The resurgence of ethnic and nationality claims may expand the basis for democracy by providing for adequate representation and devolution, but it seems that centralized unitary states are not prepared to yield at all except through contention and violence. In this sense a major challenge to the global expansion of democracy is the resurgence of ethnicity and religious fundamentalism. Both these visions still have a capability to challenge democracy from below through the articulation of deeply rooted metaphors and needs; in some instances these forces may capture state power. But these assertions may be counter-balanced by other factors, such as a sufficiently large middle class or a diffused professional cadre committed to stability and secularism. Still, we should be cautious with regard to the general enthusiasm for democracy. Democratic institutions themselves can become co-opted and used by anti-democratic forces. All too often, elections are held, but with tremendous undue influence, corruption and violence. Asian governments are very adept at such manoeuvres. This point is highlighted by Rajni Kothari as follows:

> Even less clearly recognized is the fact that the very structures that had been conceived for promoting the democratic process and providing liberation from traditional constraints – political parties, representative institutions, the judiciary – are becoming vulnerable to the influence of anti-democratic forces and are in any case proving incapable of dealing with them. Perhaps the least recognized of all are deeper forces of erosion, uncertainty and anomie that are taking hold of the mass mind at a time when growing vacuums created by the undermining of institutions and the decline of the democratic temper are being filled by specialists in violence, corruption, private arms trade and gang warfare. The sharper decline in the role of the State as mediator in social conflicts and the growing loss of faith in the political process among both the 'operators of the system' and the people at large are producing conditions of not just political instability but of incipient breakdown of the social order. The result is large social violence, the rise of negativist identities (communal and otherwise) and doctrines of exclusion and dispensability according to which entire populations are looked upon as undesirable and unwanted. (Kothari, 1988, p. 169)

11.2.4   The Crisis of Social Movements

The consequence of the collapse of the Soviet socialist utopia, as officially defined, also necessitates reformulating the goals of secular movements and the forms in which these movements will develop in the future. The working class has lost its anchor and its vanguard role, and this has led to a crisis within social movements. And it is this crisis in social movements and this lack of cohesion which has enabled fundamentalist and ethno-populist movements to capture the space for politics and transformation. The counter to the anomic conditions of modernity – the fragmentation of visions, and the revolution in time which modernity has created through the contraction of global space – is to seek solace and refuge in ethnicity and religion. Modern consumerism as one instrument of modernism may not only create heightened expectations but also lead to a greater sense of deprivation. On the other hand, there is enormous scope for popular movements, emergent networks and coalitions, which have shown a great capability to transform regimes. Many popular movements for democracy, with local and international networks forming part of a visible coalition, hold much promise.

This necessarily brief survey of modern conflicts has been intended to provide a backdrop against which we can seek meaning and conceptual clarity with regard to the violent conflicts of today. The fundamental issue is whether space exists or can be created for democratic development within many of the regions in focus here.

11.2.5   South Asia

South Asia is a specific category with some common features. These common features, however, should not tempt us to make facile generalizations. The region – geographically vast, with over one and half billion people, and a multiplicity of religions, nations, linguistic and ethnic groups – has no parallel in any other part of the world. There is futility in attempting any form of forecasting. This mosaic of ethnic and linguistic formations has inspired several religions and has been a home for many others. Hinduism, Buddhism and Jainism as religious missions compete with Islam and Christianity. But there are also features which are distinctive to the region. One of them is the size and central location of India within the subcontinent. Its sheer expanse and proximity to ethnic groups amongst its neighbours necessitate not only a high degree of sensitivity to conflicts outside its borders but a dual-track policy of managing and containing them. Moreover, India is a secular state, as is clearly and unambiguously stated in its constitution, in contrast to the theocratic states of Pakistan, Bangladesh and Sri Lanka. And unlike all its unitary neighbours, India has inherited a quasi-federal state structure.

11.3   Types of Conflicts

Within this overall conceptual framework, let us now proceed to the salience of violent conflicts within the region. It is through delimiting the specificity of these conflicts that we can discuss their resolution. I would suggest that within the framework of renewed democracy, it is likely that we will witness several types of conflicts over the next years:[5]

- Interstate conflicts
- Governance and authority conflicts
- Ideological conflicts
- Identity conflicts

This typology is proposed as a point of departure. A typology is a way of grouping instances of conflict so that common characteristics and systematic differences are revealed. But this is only a statement of purpose. Similarity and differences are cultural constructs. Typologies derive from theories – they are tools, rather than verities. To choose a conflict typology is to choose one way of looking at the world of conflict, but it need not exclude other ways. The value of a typology will only appear in its working. To ask whether it is 'true' is a semantic mismatch. You might as well ask for the 'true' colour of conflict – red or blue?

11.3.1 Inter-State Conflicts

In the South Asian region, interstate conflicts can stem from many sources – cross-border issues, including resource competition over minerals, the control of water and rivers may be significant factors in the future. Defence expenditures of all the countries in the region have increased phenomenally in recent years. At least two of the states have acquired nuclear and chemical weapons capabilities. Whilst there may not be open warfare between the states, there are many low-intensity operations and considerable war-preparedness. The successes of the high-technology war conducted against Iraq in 1991 whetted appetites for new weapons.

India and Pakistan have been in a state of war over Kashmir ever since 1947. Pakistan, controlling one-third of Kashmir, claims the entire region on the grounds that the majority of the population is Muslim. The permanent presence of UN observers in Kashmir does not prevent fighting, it merely ensures that it gets reported. Tensions at other points along the Indian–Pakistan border, as well as political problems, manifest themselves in Kashmir. It is as if the two countries express their resentment towards each other in the lofty mountains of the Karakoram Range and on the Saichen Glacier. No compromise seems to be possible between the intransigent national leaders in New Delhi and Islamabad.

The distinction between internal and interstate conflicts becomes blurred when we consider the ongoing internal struggle for self-determination by the people of Kashmir. This struggle takes place in the great Kashmir Valley and especially in Srinagar, the summer capital. Whilst Muslim Kashmiris demonstrate and fight for Kashmir to be wholly governed by Pakistan, other militants, both Muslim and Hindu, seek total independence for Kashmir.

The struggle for the right of self-determination of the Kashmiri people remains a great catalyst for war and conflagration between Pakistan and India, with other regional actors such as China and the Islamic world becoming entangled. Secular India insists on the unity and integrity of the Indian state, claiming that independence for Kashmir would provoke a full-scale Hindu backlash against the one hundred million Muslims living in India. The atrocities and human rights violations by the Indian Army have been horrendous, and India has been universally criticized for its reign of terror against the people of Kashmir.

The solution to the Kashmir question surely lies in the people having the right to choose their destiny by plebiscite or referendum. Pakistan may in the end agree to an independent Kashmir, thereby reducing Indian fears of annexation. Although the conflict seems intractable today, there are signs that in the post-Cold War world order this issue will need to be addressed once again. The UN presence has helped to monitor the conflict, but there is a need to go beyond peace-keeping operations, to peace-making and peace-building, where new frameworks will have to be created for the orderly and regulated negotiation of this longstanding dispute.

11.3.2 *Governance and Authority Conflicts*

Governance and authority conflicts are concerned with the distribution of power and authority in society. They have to do with the expansion of civil society and democratization. Conflicts over governance revolve on popular demands for democracy and political participation. The processes and struggle for democracy can be varied and highly complex. In some instances, such as in Eastern Europe, new popular movements with a non-violent project have expanded the space for civil governance. In other instances, there may be no popular forces or any significant non-governmental sector which can take over. In some cases, there may be no tradition of governance, despite the political will to introduce democracy. In others, there may not be alternatives to the existing state and government. In the South Asian region we find a complex range of issues which may be subsumed under this heading. Primarily, they have to do with due process in holding regular elections, in ensuring efficient administration, the devolution of power to municipalities or regions, and proper and orderly procedures for regime changes. Institutions and capabilities to deal with violence are also needed.

South Asia is a striking example of the erection of a facade of formal democratic theory and democracy's blatant abuse in practice. At one point, the wave of democratic change was indeed all-embracing – as in Nepal, Bangladesh and Pakistan, where for the first time popular political movements were able to wrest power from authoritarian regimes. Sri Lanka, however, has always boasted a formal civil government. The first casualty of the democratic process was Pakistan, where the administration of Benazir Bhutto was ousted by presidential order. Bangladesh experienced the overthrow of a long-standing military regime by active civic demonstration and protest. Similarly, Nepal was able to sustain a campaign for democracy which culminated in elections and the drafting of a new constitution. Sri Lanka, on the other hand, has gone through the fiction of elections and democracy, whilst encouraging extra-legal remedies to protest and dissent such as extra-judicial killings, disappearances, torture and detention.

11.3.3 Ideological Conflicts

By ideological conflicts I specifically refer to class-based conflicts with an anti-capitalist programme and an expressed manifesto intended to transform society through social revolution. These movements have recently experienced a tremendous setback, one from which they may never recover. The disintegration or stagnation of these movements occurred even before the collapse of Soviet hegemony. Today the new ideological offensive comes from the Right, with its prescription of the free market. This does not mean that ideological movements will not return to the stage to capture the social space or that new forms of organizations or new types of weapons will not emerge. Sri Lanka's Janatha Vimukthi Peramuna, the Liberation Tigers of Tamil Eelam, or the Sendero Luminoso of Peru have not vanished with the end of the Cold War. These organizations share specific characteristics, particularly in their mode of organization, tactics adopted, age composition, the basis of recruitment and in how they define armed conflict. A distinctive feature of such movements is their use of terrorism to eliminate democratic forces. They often are linked to drug and criminal networks.

11.3.4   Identity Conflicts

Identity conflicts are the most pervasive and also the most violent.

> Identity has been defined as an abiding sense of selfhood the core of which makes life predictable to an individual. To have no ability to anticipate events is essentially to experience terror. Identity is conceived of as more than a psychological sense of self; it encompasses a sense that one is safe in the world physically, psychologically, socially, even spiritually. Events which threaten to invalidate the core sense of identity will elicit defensive responses aimed at avoiding psychic and/or physical annihilation. Identity is postulated to operate in this way not only in relation to interpersonal conflict but also in conflict between groups. (Northrup, 1989, pp. 63–64)

Ethnicity is obviously an important source of legitimation, and we need not summarize the vast literature which has emerged on the subject. What is interesting here is that modernity recreates ethnicity, and certain conditions make it a dynamic category within which ethnic boundaries can be continuously redefined, allowing for political pluralism. Politicization of ethnicity is a long historical process, and polarization does not seem to take place until a certain point has been passed. Up to this point, there will normally be opportunities for conflict resolution through compromise and accommodation. Typically, however, governments provide solutions too late, and the solutions offered are inadequate to inspire confidence.

Most ethnic or minority conflicts today have a substantial international or transnational component which is a primary source of conflict. This may be due to cross-border affiliations or the existence of diasporas which provide support for these movements. Or, members of the minority community in one state may form part of the majority community in a neighbouring state, such as the Tamils in Sri Lanka; or a minority or ethnic community may straddle borders, thus involving more than one state. Although trans-border conflicts may seem benign at a particular moment, they have a tendency to flare up and escalate rapidly. The Iraq/Kuwait conflict is only one amongst many others.

The precise conditions for ethnic conflicts will depend on many factors, and the way they can be managed will depend on ethnic stratification systems. It may be that in stratification systems where there is a larger single minority with cross-border links – as with the Tamils – the conflicts may become more intractable. However, rising population pressures, poverty and the widening gap between the elites and the poor will tend to become more urgent and more clearly articulated. The search for identity often intensifies when a people's language, culture and religion are seen as the final focus of identity in a rapidly modernizing world in which they feel alienated.

11.4   Other Sources of Conflict

11.4.1   Religion and Fundamentalism

Religion in all its manifestations continues to provide a source of identity and meaning in a turbulent and modernizing world. Religion is deeply involved with internal conflicts. Despite predictions that religion would disappear with continuing modernization, it still provides a primary source of meaning and identity to many people – whether they are Christians, Muslims, Jews, Buddhists or Hindus. This religious resurgence is of significance, and many have applied the loose concept 'religious fundamentalism' to it. Too often the label has been used to denote something unlike-minded, something dark and malevolent. Often 'fundamentalists' are defined as those who do not share in one's own concept of rationality. Judaism, Christianity and Islam and Buddhism have strong ontological components of exclusion and claims for exclusivity, and have, as such, also been strong forces of mobilization. The Christian and Islamic projects, especially, are in conflict because their goals are so similar. Spreading the Gospel to the heathen, and gathering the unbe- lievers to the one True Faith – these are commands from the Highest. Judaism also lays claim to exclusivity with its concept of the Chosen People. The faithful constitute distinct, exclusive communities that confront each other as competitors for the souls of humanity. It is this exclusiveness and the claims inherent in such a concept that are potentially conflictual.

Among the most interesting sociological explanations advanced on the rise of Islamic fundamentalism or puritanism is Ernest Gellner's view of Islamic fundamentalism in terms of world civilizations, as the transformation of the central 'great tradition' of Islam into the majoritarian folk tradition:

> It allows it to assume a triple role in affirming a continuous old identity, in redu- plicating a humiliating past and poverty, and in rejecting the foreigner. And yet it also provides a charter for purification and self-discipline.[6]

Another approach is provided by Lustick in his major comparative study of ten fundamentalist movements (1988). He understands fundamentalism as involving a view of the universe and a discourse about the nature of truth that encompasses and transcends the religious domain. Thus every movement or cause becomes potentially fundamentalist.

The point about these definitions is that fundamentalist movements are seen as political movements, which may eventually involve the control or exercise of state power. They are visions which provide for an egalitarian or equal society for mankind.

Recent anthropological studies have also contributed to our understanding of the role of culture in legitimizing violence. An interesting thesis relevant to cultural sources of violence is the observation that some cultures may attribute demonic significance to other cultures and religions. It is suggested that in many ethnic or identity conflicts, ritual and religion tend to reproduce the demonic element and stigmatization of the 'other'. Other authors have emphasized the role of scapegoating in role conflicts. In extreme situations cultural scapegoating mechanisms can be used to restore societal equilibrium. Scholarly attention is also being focused on the role of 'self-fulfilling prophesies' in engendering and reproducing violence. Scapegoating can be reproduced in massive collective violence against other communities (see Kapferer, 1988).

This is not the place to discuss the ramifications of religious extremism with its periodic blood-letting in the subcontinent. However, we cannot ignore the great religious convulsions which continue to affect the various nations there.

The partition of India and Pakistan was itself a result of religious differences and interpretations between the Hindu majority and Muslims. The attack on the Ayodhya temple by Hindu extremists, the resurgence of Hindu militancy and its mobilization by political forces in India within the opposition Bharatiya Janata Party (BJP) demonstrate the fragility and strengths of secular traditions in India: fragility, in the sense of the inability of the Indian state to evolve a strategy which can contain Hindu extremism; and strength in that there is a sizeable middle class which may respond decisively and contain the violence. The Ayodhya incident will continue to haunt the government of India, which will have to develop a strategy to delegitimize Hindu extremism, so that it does not capture centre stage in Indian politics.

As mentioned earlier, only India is constitutionally a secular state within the subcontinent. The theocratic states – Pakistan, Bangladesh and Sri Lanka – can, and do, resort to religion as a means of political mobilization, a process which alienates religious minorities and creates deep divisions detrimental to the process of modernization. This conflict is amply demonstrated by the situation in Sri Lanka, where the 'Buddhist state' has to develop and obtain allegiance from a multi-ethnic and multi-religious population.

11.4.2   Violence and Militarization

The transition to civilian governments in many of the Asian societies is highly problematic and fragile. The armed forces often retain significant autonomous power, and endemic forms of violence, such as police brutality, continue. Throughout South Asia civil rights are abused, discrimination is rampant, and economic and social conditions continue to deteriorate rather than improve. In countries such as Sri Lanka, we note that grave violations of human rights continue under *civilian governments*. Severe violations of human rights, such as disappearances and assassinations, are more prominent in formal *political democracies* than under military dictatorships.

Every country in the region has experienced massive collective violence, in the form of pogroms, extrajudicial murders and guerrilla terrorism. The scale of violence over the years has been unprecedented, accompanied by a deteriorating of the standards of accountability of all sides.

Gradually there has been a major shift in thinking on political violence and terrorism. No longer is terrorism a residual category: it has now entered the mainstream of political discourse. Bombings and mass killings become commonplace, and the mass media feed the populace with day-to-day stories of gruesome killings and violence. The commonplace character of the violence projected by the media makes the populace feel powerless to act and leads them to passivity. The middle class generally ignores the effects or consequences, as long as the violence stays outside the city. The propaganda of violence portrays state terrorism as the only defensive weapon available – which leads to a general acceptance of the establishment of a national security state.

11.5 Resolving Conflicts

11.5.1 Governance and Conflict Transformation

By now I think I have made it clear that the conflict resolution stratagems which have emanated from the discourse of rationality have only partial applicability to protracted social conflicts. We may, for example, identify the conflict process in several phases – conflict formation, conflict escalation, conflict endurance, conflict improvement and conflict termination. Each phase in a conflict calls for a different type of intervention. Table 11.1 sets out the various phases in a conflict process.

Table 11.1 The Various Phases in a Conflict Process

Conflict formation	Early warning
Conflict escalation	Crisis intervention
Conflict endurance	Empowerment and mediation
Conflict improvement	Negotiation/problem-solving
Conflict transformation	New institutions and projects

11.5.2 When Conflicts Begin: Early Warning

In South Asia, as in so many other regions, there is very little recognition of the need for early warning indicators for conflict management or resolution. As yet, there are no agencies which monitor potential conflicts, except for the national intelligence services, and they are notorious for their bias and lack of credibility. There is no public agency which can work towards conflict prevention, and no ombudsmen or other governmental institutions which may facilitate preventive action. These observations hold true for the region as a whole. States tend to respond to conflicts as they arise, recognizing political power only if it is sufficiently organized. Non-governmental bodies are mostly concerned with the results of violence and are involved in humanitarian work – caring for refugees, displaced people and the casualties of conflict. Whilst there is a very large network of scholars within the region aware of the conflict situations, academic disciplines are not oriented towards action or policy. The challenge, therefore, is for existing scholarly networks now cooperating within the region to create fora for exchanging findings and views on new conflict dynamics. Such networks will need to develop linkages with non-governmental bodies, so that preventive actions may be placed on the agenda.

11.5.3   Conflict Escalation and Crisis Intervention

Very little is done to intervene when a conflict escalates into bloodshed and violence. In South Asia, recent riots and pogroms against minorities have shown clearly that the state itself is partially involved in encouraging these pogroms, and scant effort is made to halt violence or prevent it. There are no serious attempts to investigate the crimes committed against civilians and to hold law enforcement agencies accountable. Non-governmental organizations, humanitarian organizations and citizens' groups do play a role in providing relief to victims, however. Over the years, human rights organizations have developed a competence in monitoring violations by the state and have developed international links with like-minded bodies. But, generally, non-governmental organizations find themselves rendered passive or paralysed by states of emergency and violence.

11.5.4   Conflict Endurance: Empowerment and Mediation

When the conflict has matured sufficiently and both sides have managed to demonstrate their claims, either by violence or through mass pressure, then concessions are made, leading to mediation. Generally, as far as the state is concerned, this takes the form of accords, round-table conferences, pacts and agreements. Recent accords, however, do not provide grounds for any optimism. Rather than resolving conflicts, accords can serve to create new conflicts. Accords are not an attempt to bring all the parties to a consensus: they mean the exercise of power and the imposition of the will of the state. The Indo-Sri Lanka Accord is only one such example among many. Non-governmental bodies, particularly civil rights bodies, may intervene in the conflict process and some agencies may work towards mediation, but they are generally powerless.

11.5.5   Conflict Improvement: Negotiation/Problem-Solving

There are instances when negotiations begin in earnest between protagonists. But even these ceasefires and negotiations tend to break down, for a variety of reasons. In general, there is too much secrecy and a dearth of professional negotiators. Ceasefires simply become interludes used for regrouping armed forces.

11.5.6   Conflict Transformation: New Institutions and Projects

Conflict transformation is a phase in which popular forces are able to change the balance of power and actually change the regime in power. Transformation can be meaningful only if it is not a mere power transfer, but if structural changes are achieved within the society and new, effective institutions emerge.

Let us now try to classify the solutions which have been offered with regard to the kinds of conflicts discussed earlier. We may suggest:

- A high degree of regional autonomy for a minority which has already a strong territorial claim.
- Fundamental social reforms: land reform, labour rights, social redistribution of wealth etc.
- Political democracy with a free press, multi-party system, with civil and political rights.
- Consociational democracy: more complex social contracts that combine universal political rights with special provisions for vulnerable groups.

The rationalist formula may be able to deal with some of the phases in the conflict process, but not all. What is crucial to the way in which the conflict is transformed may be the timing of the various interventions and the nature of the actor who actually intervenes.

The challenge is to determine how best to utilize the democratic space available and how to transform conflicts through collective forms of non-violence, even if combatants and criminals dominate the existing democratic space. There is much to learn from popular movements in Europe and more to learn about non-violent social transformation.

11.5.7   Conflict Transformation

In the literature and theoretical discussion on conflict resolution more attention should be focused on conflict transformation. Conflict transformation is an approach which attempts to empower all the parties to a conflict, including the often passive victims. It recognizes that social conflicts need to be transformed in a less violent way – not because violence cannot achieve limited objectives, but because contemporary violence and its manifestations maim and injure all sides, including large numbers of civilians. I started this chapter by noting the inadequacies of rationality-based approaches to conflict. Let us now reflect on some comparative approaches to violent conflict. I would suggest that *each specific culture has enough resources* within itself to resolve its own conflicts. The task becomes to identify these meaningfully. Let us first take religious revivalism and fundamentalism.

In the religious discourse, religious fundamentalists tend to capture the space available. They define the terrain of the discourse, the symbols to be used and who the enemies are. Although the process of criticism and deconstruction has advanced significantly within the intellectual tradition in South Asia, unfortunately this scholarly discourse has not yet touched the mainstream religious communities. The reason for this may be that so-called progressives or modernists have not given enough thought to ways of encouraging the modernization of religion. Perhaps they have abandoned the terrain altogether. And yet, the religious discourse can only be met from within its own traditions. It is here that more needs to be done.

Another significant aspect regarding religious fundamentalism is that in many cases the priest still plays a significant role in religious interpretation. Benedict Anderson (1983) suggests that with print capitalism the monopoly of religious interpretation was no longer the exclusive privilege of the priesthood. This may hold true for Christianity. But what of other religions? What can we say sociologically about the role of priests? Generally, they are at a cross-roads, both as priests but also as institutional actors. They have to rely on meagre state stipends for support and have scant economic means.

Religious institutions need to be studied and ways found for their modernization. For example, the Buddhist Sangha is, as Professor Nathan Katz (1988) suggests:

> . . . capable of enormous political influence . . . [m]any observers have remarked that were the Sangha to rise forcefully and unequivocally against the recurrent anti-Tamil violence, then it would be stopped. Perhaps, they are the only force in Sri Lanka with the influence to effect such a response.

For a country like Sri Lanka this is highly significant. If the Sangha could develop its own approach to reconciliation and a vision of its own role in the national peace process, this would be a major force for transforming the conflict. Such developments must be rooted in the Sangha's own tradition and expressed in their own language. It would be acceptable and legitimate only if it came from within Buddhism itself, as an internal response to the crisis.

The second question I would like to address concerns conflict transformation under conditions of massive violence against civilians. Most of the countries in the region have experienced armed conflicts which have tended to encourage civilian passivity. Armed combatants define the terrain, and both the guerrilla and the security forces perpetrate the most horrendous crimes, reducing the civilian population to passivity. If we are discussing the transformation of violence and the role of civilian space and democracy, every effort must be made to recover the space and the terrain from the armed combatants. This means insisting on accountability to civilian rule for their actions. This is not easy, but there is a growing body of international norms on human rights and monitoring which is significant. The international community may play a role in providing protection. What is certain is that accountability must be ensured and sustained primarily from within the society. Furthermore, accountability must be applied to both sides, the guerrilla and state security agencies. At this point, I would like to suggest some examples from other regions of armed conflicts which may be useful for our consideration.

11.5.8 *The Democratic Revolution in the Philippines*

The democratic revolution in the overthrow of President Marcos has some distinctive features worth noting. The democratic transformation was achieved essentially through non-violent means. There was no political party which played a vanguard role but a web of popular organizations and networks which were able to intervene using techniques of non-violence. After the over-throw of Marcos and during President Aquino's period, violence continued to be endemic in Filipino society through the activities of armed guerrillas, the military and death squads. As a way of countering this violence, popular organizations developed innovative methods of expanding civilian governance. In some instances they have declared peace zones in selected local communities and called upon combatants to respect the areas as peace zones. The armed groups have, indeed, learned to respect the civilian space (see Garcia, 1992).

Another example is drawn from Colombia. A number of actors are involved – drug barons, guerrillas, the security forces and death squads. Here – with the help of urban intellectuals – 'campesinos' were able to declare their entire rural region a no-war zone. Early in 1989, when I visited the country, I had heard about an extraordinary 'self-made peace case'. In La India, a rural area rich in natural resources but inhabited by poor colonist-peasants, the armed groups (left-wing guerrillas, the army and paramilitary groups), had the civilian population trapped in the cross-fire. The peasants seemed to have only three options after 15 years of terror and killings: to leave their land, to join one of the armed groups, or to die. But in mid-1987 they decided to organize themselves and to begin a dialogue with the oppressors. They scheduled a meeting with the guerrillas to explain to them how the community could not bear the hostilities any longer. These talks took about six hours. As a result, the guerrillas promised not to oblige the community to help them any more. A few days later, the peasants talked to the army and elicited the same agreement from them. With the right-wing death squads they had non-formal talks. The three leaders from La India we worked with from 1989 were later assassinated, but the peace process continues. One consequence is that the 1990 Right Livelihood Award (the alternative Nobel Prize for peoples' organizations), was given to the people of La India.

There are many ways in which a culture of negotiations can be developed within a violent environment. These are but two examples of types of concerted actions which may transform violence.

11.5.9 Regional Frameworks for Conflict Resolution

The fate of the new world order will depend on the standards which are being set to manage and regulate so-called internal conflicts, which are largely a result of issues of self-determination. Some of the most intractable issues such as the Kashmir dispute or the Palestinian issue may well have to be settled within an international setting. The next phase of the world order may be the development of regional frameworks for economic cooperation and dispute resolution. The European Union (EU) is one model by which nation-states have invested some of their sovereignty in a supranational body. In such a framework, conflict which seemed intractable may appear tractable and manageable, such as the Basque conflict and even the issue of Northern Ireland. The Conference on Security and Cooperation in Europe (CSCE – now

OSCE), is still very fragile, but it has demonstrated a willingness to address conflicts such as that in the former Yugoslavia and build preventive programmes in Macedonia. There is a recognition that frameworks need to be created for monitoring conflicts, for creating mechanisms for conflict prevention. What may prove significant within this process is the decision to appoint an OSCE High Commissioner for Minorities to hold states accountable for minority grievances. The good offices of the High Commissioner for Minorities have been instrumental in developing effective dialogue between governments and parties in many nations troubled by ethnic minority conflicts. Another innovative mechanism being developed within the OSCE process concerns early warning and conflict prevention. The state bodies of the OSCE and the European Union exist in an environment with no shortage of complementary and competing visions, with many non-governmental networks spanning the continent of Europe, and with peace, human rights and citizens' movements with trans-border links, holding governments accountable and developing a culture of citizenship and governance. These are examples which South Asia cannot ignore.

11.6   Conclusion

Will the situation in South Asia lead to the positive developments we see within the European Union, or will South Asia witness a continuing process of fragmentation and anarchy? Another way of posing this question is to ask whether the region will be destabilized by ethnic and internal conflicts similar to what is happening in the former Soviet Union or parts of Eastern Europe. With regard to South Asia, any answer must be speculative. The recent escalation of violence in India and the challenge to the secular state by extremism may have wide-ranging repercussions.

At least South Asian governments have evolved the regional framework known as the South Asian Association for Regional Co-operation (SAARC) over time, which meets periodically to discuss and sometimes come to agreement on issues of common interest. Furthermore, economic developments in South Asia have been positive when compared to some other regions in the world. Although the incidence of poverty is growing, so too is a middle class tied to secular values. There has also been a coming together of non-governmental groups and other professional organizations, in some instances to monitor human rights or elections. However, certain parts of the region are still paralysed by armed conflict and violence. There is a need to develop a new concept of security based on more than the concept of state security: one which can guarantee the security of individuals and peoples. Such a concept should provide the foundations for stratagems which can link popular movements, render violence illegitimate through the articulation of the sovereignty of citizens and expand citizen-based forms of governance.

Notes

1. For a good discussion of conflict theory see Wehr, 1988.
2. Wallensteen & Axell, 1994. See also *SIPRI Yearbook*, especially 1988, 1989, 1990. Published by Oxford University Press. For 1989, SIPRI recorded 32 conflicts.
3. Message to Congress, 8 January 1918, President Woodrow Wilson.
4. Message to Congress, 11 February 1918, President Woodrow Wilson.
5. I have excluded resource-based conflicts and environmentally induced conflicts. These are sometimes interstate conflicts, but increasingly communities are defining the right to forest and ecologically secure zones as a collective right. These conflicts will assume greater intensity in the future.
6. Quoted in Hyman, 1985.

References

Anderson, Benedict, 1983. *Imagined Communities, Reflections on the Origin and Spread of Nationalism*. London: Verso.

Das, Veena, ed., 1990. *Communities, Riots and Survivors in South Asia*. Delhi: Oxford University Press.

Garcia, E. ed., 1992. 'Empowering People for Peace: the Philippine Experience', pp. 65–80 in Kumar Rupesinghe, ed. *Internal Conflicts and their Resolution*. London: Macmillan.

Hyman, Anthony, 1985. *Muslim Fundamentalism*. London: Institute for the Study of Conflict.

Kapferer, Bruce, 1988. *Legends of People, Myths of State*. Washington/London: Smithsonian Institution Press.

Katz, Nathan, 1988. 'Sri Lankan Monks on Ethnicity and Nationalism', pp. 138–152 in K.M. de Silva et al., eds. *Ethnic Conflict in Buddhist Societies: Sri Lanka, Siam and Burma*. London: Pinter.

Kothari, Rajni, 1988. *Transformation and Survival. In Search of Humane World Order*. Delhi: Ajanta Publications.

Lindgren, Karin, G. Kenneth Wilson, Peter Wallensteen & Kjell-Åke Nordquist, 1990. 'Major Armed Conflicts in 1989', in *SIPRI Yearbook 1990*. Oxford: Oxford University Press.

Lustick, Ian S., 1988. *For the Land and the Lord: Jewish Fundamentalism in Israel*. New York: Council on Foreign Relations.

Northrup, Terrel A., 1989. 'Dynamics of Identity in Personal and Social Conflict', pp. 55–82 in Louise Kriesberg, Terrel A. Northrup & Stuart J. Thorson, eds. *Intractable Conflicts and their Transformation*. Syracuse, NJ: Syracuse University Press.

Wallensteen, Peter & Karin Axell, 1994. 'Conflict Resolution and the End of the Cold War, 1889–1993', *Journal for Peace Research*, vol. 31, no. 3, pp. 333–349.

Wehr, Paul, 1988. *Conflict Regulation*. Boulder, CO: Westview Press.

Notes on the Contributors

SUMANTA BANERJEE grew up in Calcutta and worked as a newspaper reporter there, and later in New Delhi. Based in New Delhi now as a freelance journalist, he writes a fortnightly column for the Indian daily *The Independent*. His areas of specialization are Left politics, human rights and 19th century Bengali society and culture. His books include *India's Simmering Revolution: the Naxalite Uprising*, 1984, and *The Parlour and the Streets: Elite and Popular Culture in Nineteenth Century Calcutta*, 1989.

KUMAR DAVID is Reader in Electrical Engineering at the Hong Kong Polytechnic, with specialist research interests in computer methods in electrical power engineering. He has over 50 papers published in international journals or presented at international conferences. He also researches the politics of ethnicity and has published several articles and co-edited a book on this topic.

MEGHNA GUHATHAKURTA is Assistant Professor in the Department of International Relations, University of Dhaka, Bangladesh. She is Research Associate at the Centre for Social Studies, Dhaka, and also Associate Editor of the *Journal of Social Studies* and *Samaj Nirikkon* (quarterly Bengali journal) published from the Centre. Her publications include the book *Politics of British Aid Policy Formation toward Bangladesh*, and various articles on women's studies, aid politics and politics in South Asia.

AKMAL HUSSAIN is a development economist with a PhD from the University of Sussex and Master's degree in Economics from Cambridge University. He has been a lecturer in economics at the University of California, Riverside, and Chairman, Public Administration Department, Punjab University, Lahore. He has been a member of official committees on economic policy, including the Agriculture Census Advisory Committee, and is the author of several books and articles/papers in academic journals.

SHIREEN M. MAZARI is chairperson, Department of Defence and Strategic Studies at Quaid-i-Azam University, Islamabad, Pakistan. She holds a PhD from Columbia University, New York. She is the author of a large number of papers and articles as well as a commentator (press and TV) on contemporary strategic issues.

KHAWAR MUMTAZ is a coordinator of the Lahore branch of Shirkat Gah Women's Resource Centre with a Master's degree in International Relations from the University of Karachi. She has been a Senior Research Fellow at the Centre for South Asian Studies, the University of the Punjab, Lahore, and Assistant Editor of the weekly, *Viewpoint*. She is co-author of the award-winning book *Women of Pakistan: Two Steps Forward, One Step Back?* (London: Zed Press; Lahore: Vanguard Books). She has also published numerous articles and research papers on women's issues and politics, foreign policy and related subjects.

DEV NATHAN comes from the Left movement in India. An economist by training, he is a columnist with *Economic and Political Weekly,* Bombay. At the time of writing this chapter he was an ICSSR Senior Fellow at the Nehru Memorial Museum and Library, New Delhi. He is co-author of *Gender and Tribe,* published by Kali for Women, New Delhi and Zed Press, London in 1991. He is currently working on *Political Economy of Indian Industrialization*, to be published by Earthscan, London.

ABBAS RASHID is a freelance journalist. He has a Master's degree in International Relations from Columbia University, New York. Previously, he was contributing editor of *The Muslim,* an Islamabad-based English daily. He is currently working on a UNRISD research project on ethnic conflict and development in Pakistan.

KUMAR RUPESINGHE holds a PhD in Sociology from City University, London and a BA (honours) from the London School of Economics. He is Senior Researcher at the International Peace Research Institute, Oslo (PRIO); Secretary General of International Alert, 1992–; Chair of the International Peace Research Association's Commission on Internal Conflicts and their Resolution (ICON); and Coordinator of the United Nations University programme on governance and conflict resolution. He has published and edited many articles and books, including: *Conflict Resolution in Uganda* (James Curry Ltd, London, 1989) and *Ethnic Conflicts and Human Rights, a Comparative Perspective* (United Nations University, 1989); and a three-volume ICON book series published by Macmillan, 1992.

TANIKA SARKAR graduated from Calcutta University in 1974 and received her doctorate from Delhi University in 1982. She has written several articles on various aspects of the national movement in Bengal and teaches history at St Stephen's College, which is a part of Delhi University.

JAYADEVA UYANGODA, BA (Sri Lanka), PhD (Hawaii), is Senior Lecturer in Political Science and Co-director, Centre for Policy Research and Analysis, University of Colombo, and a human rights and peace activist.

Name Index

Subject Index